T0211254

CLEAN ROOM DESIGN

Minimizing Contamination Through Proper Design

■ ■ ■

Bengt Ljungqvist
Berit Reinmüller

CRC Press
Taylor & Francis Group
Boca Raton London New York

CRC Press is an imprint of the
Taylor & Francis Group, an **informa** business

CRC Press
Taylor & Francis Group
6000 Broken Sound Parkway NW, Suite 300
Boca Raton, FL 33487-2742

First issued in paperback 2019

© 2002 by Taylor & Francis Group, LLC
CRC Press is an imprint of Taylor & Francis Group, an Informa business

No claim to original U.S. Government works

ISBN-13: 978-1-57491-032-2 (hbk)
ISBN-13: 978-0-367-40114-6 (pbk)
Library of Congress Card Number 96-43301

Library of Congress Cataloging-in-Publication Data

Ljungqvist, Bengt
    Clean room design : minimizing contamination through proper design.
    Bengt Ljungqvist, Berit Reinmüller.
       p. cm. .
    Includes bibliographical references and index.
    ISBN 1-57491-032-9
    1. Clean rooms. I. Reinmüller, Berit, II. Title
TH7694.L57   1996
620.8—dc20   96-43301

Visit the Taylor & Francis Web site at
http://www.taylorandfrancis.com

and the CRC Press Web site at
http://www.crcpress.com

CONTENTS

PREFACE

The work presented here is partly based on publications from international congresses, the *PDA Journal of Pharmaceutical Science and Technology*, and the *European Journal of Parenteral Sciences*. This work is based on cooperation in the field of Safety Ventilation between Building Services Engineering, Kungl. Tekniska Högskolan, KTH (Royal Insititue of Technology), and Pharmacia & Upjohn, both in Stockholm, Sweden; the work has been carried out over the period 1987–1996.

However, this is the first time the whole work has been presented in its entirety, in order to accommodate requests and educational purposes. The authors, therefore, wish to take this opportunity to extend their sincere thanks to all who have supported this work.

As far as the authors are aware, it was Dr. Claes Allander, Professor Emeritus, KTH, who first realized the importance of closed streamlines in the contamination control context, and this work may be regarded as a continuation of Professor Allander's ideas.

The authors wish to thank Mr. Mats Johansson, M. Sc., Manager of Microbiological Technique at Pharmacia, who gave the necessary support to accomplish this work; Mr. Hugo Thelin, former Manager of Astra and of Kabi Pharmacia, for his moral support and never failing visions of the necessity of new knowledge and in improvements in production safety; Professor Ove Söderström, KTH, for valuable discussions about the mathematical presentation in Appendix A1, Dr. Engelbrekt Isfält, KTH, who has performed the simulation runs in Appendix A2; and Mr. Russell E. Madsen, Vice President of Scientific and Technical Affairs PDA, for encouragement and linguistic support.

The work is designed as a text for educational purposes and as a reference for practical applications.

Stockholm
October 1996

Bengt Ljungqvist
Berit Reinmüller

INTRODUCTION

This book describes the theoretical relations for the dispersal of airborne contaminants and illustrates the validity of these equations occurring during factual situations, where a number of observations on air movements in open unidirectional air flow units supplied with HEPA-filters are described.

In factual situations, the aerodynamic system, which governs the dispersion of contaminants, in reality is always very complicated so that risk situations must be mapped and assessed empirically. The presence of a person can give risk to wakes that may be stable or unstable. The unstable situations are, in most cases, caused by the influence of arms and hands. As part of the microbiological assessment of aseptic processes carried out in clean zones, it is important to investigate that such vortices do not occur in the clean working areas.

As the level of airborne contaminants in the operational environment may have an effect on the level of product contamination, the microbiological assessment of aseptic processes is important.

A method—the LR-Method—is described for microbiological assessment in unidirectional air flow units by using Visual Illustrative Methods, Particle Challenge Tests (measured by a particle counter), and calculation of Risk Factor for the dispersion and/or induction of particles.

Different types of weighing stations are discussed and qualification measurements with a tracer gas method are described.

1

DISPERSION OF AIRBORNE CONTAMINANTS

1.1 General

The air may move in two different ways. One of these is characterized by a smooth flow, free of any disturbances, such as small and temporary vortices or eddies. This is known as laminar flow. The other type of flow is characterized by small and temporary fluctuations caused by instabilities. The flow velocity is no longer constant but more or less fluctuates around an average value. This is known as turbulent flow and the disturbances are often interpreted as being small, temporary eddies.

In order to estimate the problems associated with the transport of contaminants by air, we must understand how this occurs. We must assume that, with traditional ventilation processes and the rules we apply, the air in the rooms is more or less turbulent.

The aim is to arrange ventilation in such a way that there is a certain basic flow of air. An organized basic flow implies that the flow can be characterized by means of stream lines, i.e., the paths taken by weightless particles in the room as they follow the air stream, if the turbulent fluctuations are ignored. The transport of contaminants due to streamline flow is often described as "convective transport".

The simplest system for an analysis of the transport of contaminants by ventilation is, therefore, convective transport along the streamlines. The disturbances caused by turbulence (turbulent diffusion) are superimposed on this. Obviously, if there is no turbulence, turbulent diffusion is replaced by molecular diffusion or Brownian motion. It can generally be assumed, in regions with well-defined air flow fields, that the settling velocity of contaminants is negligible, which implies that gravitation plays an inferior role.

With the assumption of a constant value of the diffusion coefficient, the diffusion equation in a velocity field in rectangular coordinates becomes:

$$\frac{\partial c}{\partial t} + v_x \cdot \frac{\partial c}{\partial x} + v_y \cdot \frac{\partial c}{\partial y} + v_z \cdot \frac{\partial c}{\partial z} = D \cdot \left(\frac{\partial^2 c}{\partial x^2} + \frac{\partial^2 c}{\partial y^2} + \frac{\partial^2 c}{\partial z^2} \right) \quad (1)$$

where $\quad c \qquad\qquad$ = concentration

$\qquad\qquad v_x, v_y, v_z \quad$ = velocities in x-, y-, and z-direction

$\qquad\qquad D \qquad\qquad$ = diffusion coefficient

This gives the simplest possible mathematical model that describes a system with regard to transport of contaminants emitted from a source at an arbitrary position.

1.2 Unidirectional Flow

Dispersion from a fixed source in a uniform parallel flow is described theoretically and experimentally, inter alia, by Bird et al. [1], Fuchs [2], Hinze [3], and Ljungqvist [4]. For a continuous point source in a parallel flow with constant velocity, v_o, in the x-direction, the solution of the simplified diffusion equation in the velocity field mentioned above when $\partial c/\partial t = 0$ becomes:

$$C = \frac{q}{4\pi Dr} \cdot e^{-\frac{v_o}{2D}(r-x)} \cong \frac{q}{4\pi Dx} \cdot e^{-\frac{v_o(y^2+z^2)}{4Dx}} \qquad (2)$$

where q =   outward flow from the point source

   v_0 =   constant velocity in the x-direction

   r = $(x^2 + y^2 + z^2)^{1/2}$

The case of parallel air flow has been investigated experimentally by Ljungqvist [4] for a room with a cross section of 1.7×1.7 m^2. It was found that, with a turbulence-free inlet, there was a complete absence of eddies for air velocities up to approximately 0.3 m/s. If the inlet opening of a room was equipped with a turbulence-generating grid, the degree of turbulence obtained at even relatively low air velocities (0.2 m/s) was such that the diffusion, due to the turbulence, determined the development of dispersal for both gases and particles.

Values of the turbulent diffusion coefficient were determined by Ljungqvist [4] with the aid of Equation (2) from concentration curves measured at air velocities of 0.20 m/s and 0.45 m/s. The diffusion coefficients obtained for the two velocities were approximately 1.4 cm²/s and 2.4 cm²/s, respectively.

Ljungqvist and Reinmüller [5] have estimated the "critical contamination region" in undisturbed and disturbed parallel flow fields with dispersion from a fixed contamination source. A qualitative solution in an undisturbed parallel flow is given in Figure 1, when the velocity is 45 cm/s and the distance between the working surface and the point source is 30 cm.

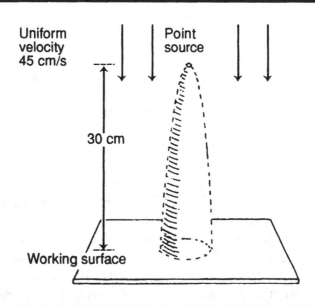

Figure 1. Critical contamination region in a uniform parallel flow field. Qualitative solution of the diffusion equation in a velocity field with the velocity 45 cm/s, diffusion coefficient 2.4 cm²/s and the distance between the working surface and the contamination source 30 cm.

1.3 Vortex

A vortex is characterized by the fact that the stream lines are closed within a region, which in the following is referred to as the *vortex region*. According to the laws of aerodynamics, tangential velocity in the vortex region should increase as the center of the vortex is approached. However, systematic investigations by Ljungqvist [4] show that this is not always the case in vortices formed in ventilated rooms. Everything indicates that the air mass within the vortex region moves as a rigid body under the influence of powerful turbulence. A certain amount of energy is, therefore, needed to maintain a vortex, and in most cases, this energy is obtained from the kinetic energy of the air on its entry into the room. The greater the kinetic energy of the air in the room, the greater the chance of vortices occurring with closed streamlines.

Owing to the fact that the streamlines are closed, there is no convective removal of contaminants emitted within the vortex region. It is only turbulent diffusion within the vortex that causes removal of the contaminants. In a room where contaminants are emitted within a vortex region, the average concentration of contaminants inside the vortex region may, for instance, be 10 times higher than in the air extracted by ventilation. Transport due to turbulent diffusion does not increase in proportion to the vortex velocity but is changed at a slower rate. The results from measurements by Ljungqvist [4] show that the turbulent diffusion coefficient is governed by an exponential law and that the value of the index is about 0.7. This is in agreement with the relation of the two values of diffusion coefficient given earlier for the turbulent unidirectional air flow. On the other hand, the absolute value of the turbulent diffusion coefficient is almost 20 times higher. This confirms the violent nature of the turbulence.

If, therefore, for a certain ventilation rate of flow there is a vortex region within which a contaminant is emitted and this rate of flow is doubled, the concentration of the contaminant in the vortex region is not halved; it drops to somewhere between the original value and half this value.

The mathematical treatment can be formulated with the problem used to solve the diffusion equation for a gas that rotates as a rigid body, because there is a continuous release of the diffusing substance from a source situated at a certain distance from the axis of rotation.

If the rotation takes place about the z-axis with the angular velocity ω and if the source is located at the point $(a, 0, 0)$ with the outward flow q, the solution is given by Ljungqvist [4] in cylindrical coordinates $(x = \rho \cos \varphi, y = \rho \sin \varphi, z)$, and the expression of the concentration becomes:

$$c = \frac{q}{8\pi^{3/2}} \int_0^\infty e^{-(\rho^2 + a^2 - 2a\rho\cos(\varphi - \omega t) + z^2)\frac{1}{4Dt}} \frac{dt}{(Dt)^{3/2}} \quad (3)$$

where a = distance between the point source and the center of vortex (origin)

       ω = angular velocity

It is natural that a certain preference should be given to the x,y-plane, as it is in this plane that the source of emission is located.

With $z = 0$ and appropriate substitutions* in Equation (3), the expression of the concentration can be interpreted as the product of the concentration at the origin and a non-dimensional factor. The solution of Equation (3) for this case gives the expression

$$c = \frac{q}{4\pi Da} \cdot F \qquad (4)$$

*where F = non-dimensional factor

The factor F is, apart from the positional coordinates, a function only of the angular velocity when the diffusion coefficient is constant, i.e., the turbulence is isotropic. For a certain angular velocity, the factor F can be plotted as the vertical coordinate perpendicular to the x,y-plane (ρ, φ-plane). In this way a surface is obtained, the general appearance of which is shown in Figure 2.

Figure 2 shows that the mean value of the concentration over the center region beside the point of emission is considerably higher than the concentration outside. The above, therefore, allows us to use the concept of contamination accumulation in the context of vortices, especially as the diffusion coefficient does not increase linearly with the air change rate. In this way, it has also been shown that the concentration of contamination in the exhaust air cannot be used for assessment of the risks in the system discussed here.

Figure 2. Model of the non-dimensional factor F^*.

* With the substitutions in Equation (3)

$$\Omega = \frac{a^2\omega}{4D} \ , \ \xi = \frac{\rho}{a} \ \text{and} \ \tau = \frac{4D}{a^2} t$$

the factor F becomes

$$F = \frac{1}{\sqrt{\pi}} \int_0^\infty e^{-\{1+\xi^2-2\xi\cos(\varphi - \Omega\tau)\} / \tau} \frac{d\tau}{\tau^{3/2}}$$

For total derivation, see Ljungqvist [4].

2

CONTAMINATION RISKS

2.1 Flux Vector

The risk of contamination does not only depend on the concentration of the contaminants, which is of critical importance, but also the motion of the contaminants. If particulate contaminants are being considered, then it is the rate of incidence of the particles that characterizes the risk. When considered mathematically, this incidence is of vector character and the term *flux vector*, or *impact vector*, is used, see, e.g., Friedlander [6], Ljungqvist [4].

The risk will, to a great extent but not entirely, be dependent on the vector

$$\boldsymbol{K} = -D \operatorname{grad} c + \boldsymbol{v} \cdot c \tag{5}$$

where c =   particle concentration

$\boldsymbol{v}$ =   velocity vector

In principle, the numerical value of K indicates the number of particles passing a supposed unit area, placed perpendicular to the direction of particle flow, per unit time.

The diffusion part in Equation (5) has importance only when situated close to the particle source in properly ventilated regions, and it may be assumed that the dominant contribution to K is due to convectively transported particles, i.e., $v \cdot c$.

The expression for the contamination risk in the main flow direction of a unidirectional air flow is easily derived from Equation (2) and is proportional to air velocity and concentration, while the expression for a vortex is more complicated.

To get a better understanding, the conditions of the vortex will, as before, be dealt with only in the x, y-plane. Supposing the point of emission is located in, or in the immediate vicinity of, the center of the vortex, a hyperbolically decreasing concentration of contaminants will be obtained. This is yielded directly by Equation (3) if a and z here are put equal to zero. The result then will be

$$c = \frac{q}{4\pi D\rho} \tag{6}$$

At the same time the velocity v, which in this case has a tangential direction and is proportional to ρ, increases. The result is, therefore, that the tangential flux vector has a constant numerical value over the entire x, y-plane. The conclusion must, therefore, be drawn that a vortex of the kind considered in this chapter is very detrimental from the point of view of contamination risk when contaminants are being released inside the vortex.

External force fields of interest in particle transport are gravitational, electrical, and thermal—the last produced by temperature gradients. The quotient between the force field and the friction coefficient provides the migration or drift velocity in the field. In a mathematical treatment, this drift velocity should be added to the velocity vector in Equation (5). For example, the friction coefficient for the gravitational field is usually based on Stokes' law, and the drift velocity becomes the settling velocity due to gravity.

Consider a chamber of height H containing, at time zero, a specified initial concentration c_0 of uniformly distributed monodisperse particles. If there is no motion of the air and the diffusion is neglected, all the particles will be settling with the same constant settling velocity v_s and the particle flux becomes $v_s \cdot c_0$. The concentration becomes zero everywhere in the chamber after a time equal to H/v_s has elapsed.

At the other extreme, when the air is completely turbulently mixing, it is assumed that the concentration is uniform throughout the chamber at all times. Diffusion and deposition on the walls are assumed to be negligible. The particle settling velocity is superimposed on the vertical components of convective velocity. Because the up and down components of convective velocity are equal, every particle will have an average net velocity equal to v_s. The concentration of particles decays exponentially with time and as such never reaches zero. The concentration reaches $1/e$ (ca 37%) of the original concentration in the same time (H/v_s) that is required for complete removal in the above given case with no motion of the air (see, e.g., Fuchs [2] and Hinds [7]).

The expression for the particle flux at the time t in this case with monodisperse aerosol is

$$K_s = v_s \cdot c_0 \cdot e^{-\frac{v_s t}{H}} \tag{7}$$

where c_0 = initial concentration

v_s = settling velocity

H = height of chamber

The number of particles of a monodisperse aerosol settling in a time t on the unit area of the bottom of the chamber becomes

$$N_s = \int_0^t K_s dt = c_0 H \left(1 - e^{-\frac{v_s t}{H}} \right) \tag{8}$$

In a horizontal unidirectional air flow, the settling velocity becomes the vertical net velocity component. If the air flow is turbulent, the diffusion is also of importance (compare with Figure 1).

It is assumed in a parallel flow field that, next to surfaces along the main flow direction, there is a thin sublayer in which the transfer of momentum is dominated by viscous forces, and the effect of weak turbulent fluctuations can be neglected. The situation is quite different for particle diffusion. In this case, even weak fluctuations in the viscous sublayer contribute significantly to transport.

In a unidirectional air flow, wakes and vortex streets are easily created behind obstacles. The air flow often leads to stagnation regions in front of machinery and working surfaces situated perpendicular to the main flow direction. The air movements in this case are mostly irregular, and in a real case situation, air movements are not so easy to predict (see chapters 3 and 4). These regions can, of course, have an impact on the contamination risk, and in regions with increased turbulence, the diffusion part in Equation (5) is also important.

2.2 Common Application of Viable Particles

Airborne particles carrying microorganisms are shed from people, and the particle size is mainly in the range of 5–20 μm. An equivalent mean diameter between 12 and 14 μm (spherical particle with the unit density of 1 g/cm³) has been established. For a more thorough description see Whyte [8], Noble et al. [9], and Clark et al. [10].

According to Stokes' law, the settling velocity for a spherical particle of 20 μm and unit density (1 g/cm³) in air is less than 2 cm/s. This velocity is much less than that of air in a unidirectional flow (30–50 cm/s), which is commonly used in pharmaceutical production. This means that the settling velocity in the main flow direction plays an inferior role compared to the unidirectional flow velocity.

A one-dimensional special case of Equation (5) occurs when the diffusion part is neglected and the velocity vector depends only on the settling velocity. When an equivalent mean diameter of bacteria-carrying particles can be established, the settling velocity becomes a constant value. If the concentration of bacteria-carrying particles in the air, the area of exposed surface, and the exposure time are known, the number of bacteria-carrying particles deposited can be calculated. When the concentration of bacteria-carrying particles is uniformly distributed and constant during exposure time, the expression for the number of particles deposited is

$$N_d = v_s \cdot c_b \cdot A_e \cdot t_e \qquad (9)$$

where c_b = concentration of bacteria-carrying particles

A_e = area of exposed surface

t_e = exposure time

With the assumption that the average size of bacteria-carrying particles is 12 μm, and hence the settling rate is 0.462 cm/s, the contamination risk will be equivalent to the contamination rate given by Whyte [8].

With the above given assumption, Equation (9) can be used when there is no motion of air (still air), for completely mixing air (fully turbulent air), and in vertical direction in the case with horizontal unidirectional air flow with little influence of turbulent fluctuations (laminar air flow). Since the demand in Equations (7), (8), and (9) of uniformly distributed concentration in the air is not always fulfilled in real case situations, other theoretical or experimental methods must be taken into consideration when estimating the contamination risks. In chapter 5 an experimental method for microbiological assessment of potential risks in clean rooms is described.

2.3 Electrostatic Charges

Contamination risks can also occur at electrostatically charged surfaces, which can increase the deposition of particles up to several orders of magnitude. Electrostatic forces can overcome the unidirectional air flows or gravity, and particles that would otherwise not have been attracted to the surface will end up on charged surfaces. Once a particle is bonded to a surface, it is extremely difficult to move.

Primary electrostatic charges are generated from friction or separation between two different materials. At the contact surfaces of the two materials, a dynamic exchange of valence electrons will occur. The quantity of charges depends on several factors, such as surface resistivity, number and size of contact points, speed of separation, and surface temperatures.

In principle, there are two main methods for neutralizing electrostatic charges. The first is to use materials with low surface resistivity, so that the materials do not charge by friction and are able to neutralize charges of the objects. The second method is to use air ionizers to neutralize electrostatic charges. This second method is used when it is not possible to use materials with low surface resistivity in the sensitive clean zones or when the product with its container has high surface resistivity.

Other methods, such as grounding, humidity control, and antistatic solutions, should also be considered, if they are possible to use, in order to reduce or eliminate static charges from the clean room environment.

2.4 Risk Estimation

It is obvious that the risk of contamination is more intimately associated with the flux vector than with concentration. It must also be pointed out that the mathematical and physical treatment of risks must be differentiated in view of their character. From a general point of view, the above reasoning can be characterized in the following example.

> Somewhere in a well-defined room, a stationary source of contaminants is situated. In the room there is some ventilation, and this is also assumed to be stationary. This situation prompts the following obvious question: What is the distribution of concentration of the contaminants for a given flow configuration of air?

This question is extremely far-reaching in its implications, and it does not seem possible even to indicate a general method whereby it can be tackled. In spite of this, its formulation is important, because, with this formulation, the problem is brought to a head. Namely, if we locate the source at a point, P1, in the room and calculate the concentration at another point, P2, then this concentration, together with the knowledge of the velocity of flow at point P2, will provide information concerning the risk of transmission from point P1 to point P2. In other words, if the points P1 and P2 can be located anywhere, the risk situation in the room can be clarified.

The above discussion shows clearly the importance of having knowledge about the interaction between air movements and dispersion of contaminants when contamination risks are assessed.

3

WAKES, FACTUAL SITUATIONS

Stable vortices, with their need for energy-demanding turbulence, are unusual. When they do occur, it is mostly in the form of wakes, which are set up behind obstacles in a high energy, more or less parallel air flow. Vortices are generally unstable, i.e., they have limited duration. Such vortices are often periodic, i.e., they are formed, decay, die out, and are formed again, and so on. The frequency need not be uniform but may vary. This is obviously the case when it is the movements of a person that give rise to a vortex.

With visual illustrative tests (Ljungqvist [4, 11]) in which the emitted contaminants are replaced by isothermal smoke and the dispersion is recorded by means of still photographs and film, it has been shown that the presence of a person can give rise to wakes that may be stable or unstable. The unstable situation is, in most cases, caused by the influence of arms and hands. Figure 3 shows a parallel flow field in which a dummy is placed to the side of a tube from which smoke is continuously released. It is obvious from the photograph that the dummy does not disturb the dispersion of the smoke.

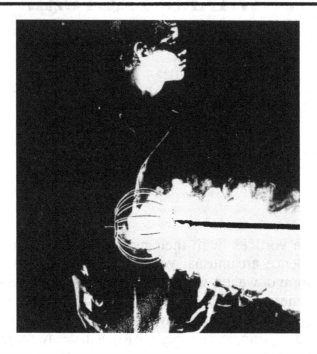

Figure 3. Dummy placed at the side of the point of smoke emission in a parallel flow field. Hand in the lower position.

However, as soon as the hand of the dummy is raised, there is a dramatic change in the dispersion configuration, as can be seen from Figure 4.

It must be emphasized that this situation is extremely unstable. If, however, a person is placed in front of the point of release of the smoke, a stable wake region is formed. This is shown in Figure 5. We can see that the smoke follows the contour of the test person's body and reaches the respiration zone. It is evident from these photographs that the release of contaminants in the lee of a person in the vortex region results in contaminants being transported to the breathing zone.

Figure 4. Dummy placed at the side of the point of smoke emission in a parallel flow field. Hand in the upper position.

Kim and Flynn [12] describe experimental results, using hot-film anemometry and flow visualization techniques, and suggest that air flow around a person immersed in a uniform, free stream has a three-dimensional nature. Above the chest level, the down-wash effect is important; from chest to elbows, a combination of the down-wash and vortex shedding exists; and from waist to hip, the vortex shedding appears dominant. In the region subject to the vortex shedding, each vortex is shed downstream at a dimensionless frequency (Strouhal number) of 0.19. The average area of the vortex is 0.7 times the area of a circle with a diameter equal to the width of the dummy.

Figure 5. Test person placed in front of the point of smoke emission in a parallel flow field.

A coherent vertical flow is present in the proximity of the body: the upper section (above hip level) where the mean flow is directed upwards, and the lower section where the mean flow is directed downwards. Additionally, the end of the reverse flow zone reaches at least two widths downstream of the worker and implies that a hand-held contaminant source cannot escape the influence of the recirculating flow. Figures 6 and 7 show the region of reverse flow and the air flow structure downstream of the dummy.

The results by Kim and Flynn [12] threw new light on the air flow around a person in a unidirectional air flow. The smoke configuration in Figure 5 shows similarity with the drawing

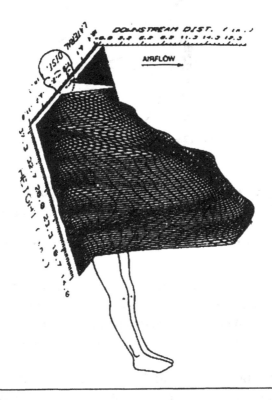

Figure 6. Region of reverse flow averaged for three different free stream velocities (From Kim and Flynn [12]. Reprinted with permission of the *American Industrial Hygiene Association Journal*.).

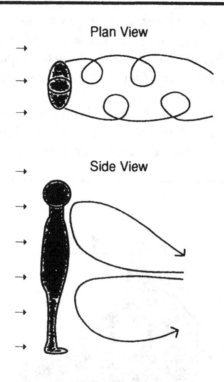

Figure 7. Air flow structure downstream of the dummy (From Kim and Flynn [12]. Reprinted with permission of *American Industrial Hygiene Association Journal.*).

in Figure 7, because the smoke emission is in the upper part of the body. Of course the configuration of the body and the convection flows may play a role in determining air patterns and exposure.

According to Clark and Edholm [13], the maximum air velocities found over the face of a person in an upright position are 0.3–0.5 m/s compared with 0.05 m/s for a person in a prone position. For a standing person, the flow over the face is that which has been developing in velocity and thickness over the entire height of the body. In contrast, when a person is prone the flows are generally slower and thinner. A sitting position produces flows that are intermediate between upright and prone.

4

OPEN, UNIDIRECTIONAL AIR FLOW BENCHES

4.1 Air Movements in Empty Benches/Units

Within pharmaceutical production, aseptic sterile processes are required to be carried out in a class 100 environment (U.S. Federal Standard 209E, "Airborne Particulate Cleanliness Classes in Cleanrooms and Clean Zones"). To avoid particle contamination in critical process areas (clean air zones), unidirectional air flow with high efficiency particulate air (HEPA)-filtered air is used. The aim is to have sterile, particle-free air flow over the open drug product and to have any particles that may be emitted swept away from the exposed drug product.

In vertical unidirectional air flow benches, the area along the vertical side walls in front of the operator is usually entirely or partly open. When the other side walls reach down to the working surface in the bench/unit a stagnation flow with stationary vortices is usually created, as shown in Figure 8.

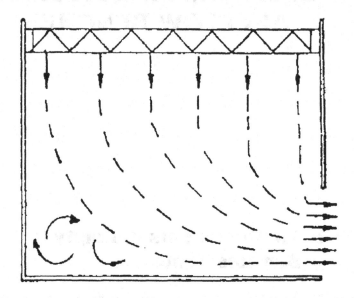

Figure 8. Vortices in a unidirectional air flow bench with covered side walls.

This vortex region can be demonstrated by using a smoke photography technique (see Figure 9). In this case when the flow can be considered as two-dimensional, the vortex region can easily be predicted by using modeling software for personal computers (see, e.g., Busnaina [14] and Busnaina et al. [15]). Such a computer-based prediction is shown in Figure 10 for a vertical unidirectional air flow bench in a mixed flow clean room with velocities of 0.45 m/s and 0.1 m/s in the bench and the room, respectively.

To avoid such vortices, the side walls must be designed with openings. At a long bench designed with equally large openings on the longitudinal and opposite wall, the flow can be considered as two dimensional. If it is further assumed that the flow may be regarded as turbulence free, it is possible

Figure 9. Vortex region in a bench with the air pattern shown in Figure 8 when air velocity is 0.45 m/s.

to find an exact solution for Navier-Stokes equations. The solution was first given in a thesis by Hiemenz [16] in 1911.

This plane flow leads to a stagnation point in the middle of the bench on the working surface. Furthermore, it is worth noting that the boundary layer is proportional to the square root of the kinematic viscosity, and that the thickness of the boundary layer does not vary along the working surface. In reality, a stagnation region often arises where the height and appearance can vary as shown in Figure 11.

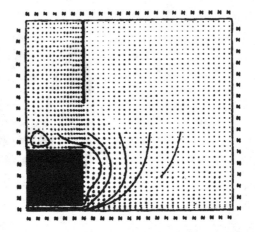

Figure 10. Computer-based prediction of the flow field (dashed lines) and contaminant particle transport (solid lines) for a vertical unidirectional air flow bench (0.45 m/s) in a mixed flow clean room (0.1 m/s).

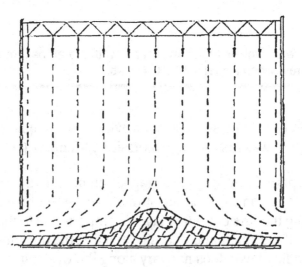

Figure 11. An example of observed stagnation region in a bench with vertical unidirectional air flow.

In the unidirectional air flow of an open bench, a vortex street is easily created behind small obstacles. Such an obstacle can be as insignificant as a small lamp or a fixture connecting HEPA filters.

Ljungqvist et al. [17] have, with help of isothermic smoke, visually depicted the air movements behind such a horizontal 30 mm wide fixture at an air velocity of 0.45 m/s. The observed flow pattern is schematically shown in Figure 12.

The flow pattern in Figure 12 has a violent turbulent region, characterized by a vortex street and two free vortices rotating in opposite directions close to the working surface. These observed vortices are known as an *irrotational* or *free vortex* and sometimes as a *potential vortex*. This vortex is characterized by the fact that the velocity varies only with the radius and increases toward the center.

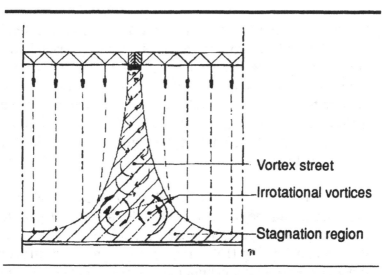

Figure 12. Observed flow pattern behind an obstacle in a vertical unidirectional air flow bench (long side).

This free (irrotational) vortex, which is often described in literature, differs conspicuously from the vortex earlier described (Ljungqvist [4]), where the velocity decreases linearly to the center. This vortex is known as a *forced* or *rotational vortex*, and because the fluid rotates like a solid body, the term *rigid body rotation* has also been used. From the standpoint of contamination, both types of vortices will accumulate contaminants.

4.2 Air Movements in Benches/ Units with Obstacles

The air movements outside the vortex region in the bench shown in Figures 8 and 9 are depicted visually with smoke at an air velocity of 0.45 m/s in Figure 13.

If a bottle is placed in the bench, a wake is easily created in the region with horizontal flow. This is depicted visually with smoke in Figure 14. If the bottle is situated close to the opening of the unit, ambient air will be entrained into the clean zone in the bench. The length of the reversed region can be estimated to 2–3 times the diameter of the bottle, and can reach twice this length when the bottle is situated just beside the side wall.

The Reynolds number, which is directly proportional to the air velocity and the size of the obstacle, is a critical quantity. According to photographs presented, inter alia, by Batchelor [18] and Schlichting [19], a regular Karman vortex street in the wake of a circular cylinder is observed only in the range of a Reynolds number from about 60 to 5000. At lower Reynolds numbers, the wake is laminar, and at higher Reynolds numbers, there is a complete turbulent mixing.

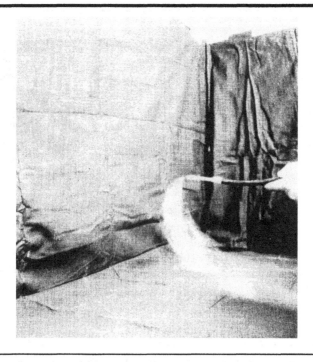

Figure 13. Undisturbed smoke dispersion in the bench shown in Figure 8 at an air velocity of 0.45 m/s.

In the case shown in Figure 14, the Reynolds number is around 2700. In the situation in Figure 12, where the arrangement between the two filter modules forms a flow obstacle, the Reynolds number becomes 900.

One should be cautious when comparing the Reynolds number from regular Karman vortex streets with the Reynolds number calculated from factual situations in clean benches, as the air flow behind an obstacle is mostly not a typically formed Karman vortex street behind an indefinitely long circular cylinder. The wake situations during actual conditions often seem to have a three-dimensional structure (compare with Figures 6 and 7).

Figure 14. Dispersion of smoke behind a bottle in the bench shown in Figure 13.

A more complicated wake situation is shown in a production unit with a vertical flow field in Figure 15. If a hand is placed over the smoke source in Figure 13, a wake region is created in the vertical flow field, as shown in Figure 16.

Essentially, vortices caused by people are of two kinds. Relatively stable and stationary wakes are created by the body. Unstable and non-stationary vortices arise as a consequence of the movements of the body. In this respect, it is obvious that the movements of the hands and arms play a significant role in creating unstable situations. Some work situations in a clean air bench with a horizontal uni-directional air flow are demonstrated by using a smoke technique (see Figures 17–19).

Figure 15. Dispersion of smoke in a vertical unidirectional flow bench showing a wake region behind obstacles.

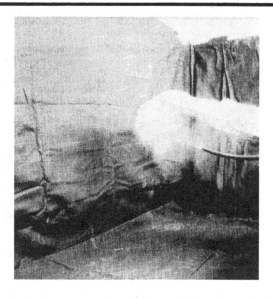

Figure 16. Dispersion of smoke behind a hand (arm) in the vertical unidirectional flow bench shown in Figure 13.

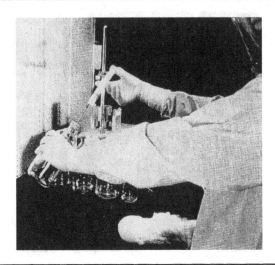

Figure 17. Dispersion of smoke in a horizontal unidirectional flow bench when operator's hands are in an upper position (undisturbed smoke dispersion).

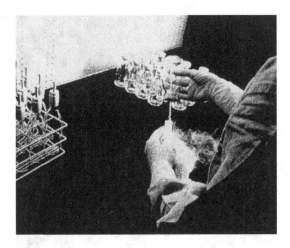

Figure 18. Dispersion of smoke in a horizontal unidirectional flow bench when operator's hands are in a lowered position (wake region behind hands).

Figure 19. Dispersion of smoke in a horizontal unidirectional flow bench when operator's hands are in motion.

5

MICROBIOLOGICAL ASSESSMENT

5.1 Supplementary Test Methods

In a clean room the microbiological burden is usually very low, and it is difficult to achieve results that are statistically significant for microbiological assessment. Air supplied to a clean room is filtered and does not usually contribute to airborne contamination. The cleanroom is additionally positively pressurized to prevent contamination from adjacent areas. Therefore, the sources of airborne particles within the room are essentially people and machinery. People are the main source of airborne bacteria.

Clean room monitoring can be performed in several ways and with the aid of different techniques. Knowing the limitations of the chosen system is important. The number of colony forming units (CFUs) detected by one method cannot be directly compared with results from another method. Results from parallel sampling—with different sampling equipment used under real conditions—are shown in Figures 20 and 21 (Ljungqvist and Reinmüller [20]). The results show

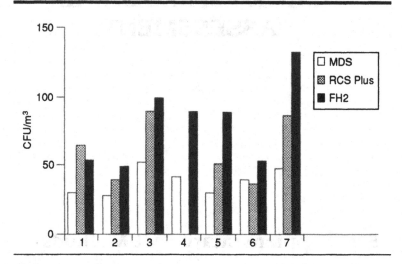

Figure 20. Observed numbers of airborne CFU from three air samplers (MD8, RCS Plus®, and FH2) during 7 parallel sampling periods.

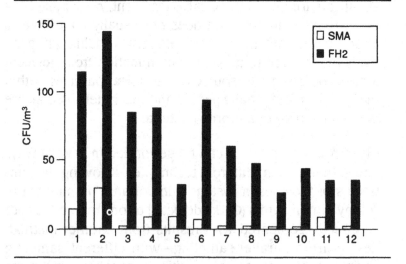

Figure 21. Observed numbers of airborne CFU from two impaction samplers (SMA and FH2) during 12 parallel sampling periods.

clearly that different sampling equipment gives different numeric values.

Sampling of airborne bacteria as CFUs often produces results of less than 1 CFU per sample. With the common microbiological methods, it is difficult to evaluate how the performance of single operations affects the microbiological risks of a process.

It is extremely difficult to evaluate potential microbiological risks, and too often, the microbiological assessment is based on poor data and subjective conclusions from visual observations. This means that the assessment sometimes could be based on an unsystematic approach, where the intuition of the responsible microbiologist plays an important role.

To achieve an objective approach to the microbiological assessment of potential risks in clean rooms, at least two methods are possible: the first is by increasing the microbiological burden, and the second is by increasing the particle level.

The first approach, which is direct and has been reported by Bradley et al. [21], uses a technique with a containment room where a suspension of Bacillus spores in water is blown out into the air around a filling machine. The results from this investigation show that the level of airborne microorganisms in the filling environment has a profound effect on the level of product contamination. Furthermore, a direct relationship was reported between the extent of product contamination and the level of airborne microorganisms when the filling equipment—a blow/fill/seal machine—was investigated. When the air shower with HEPA-filtered air over the filling zone was on, a tenfold reduction in product contamination was observed, as opposed to when the air shower was off. This allows, according to the authors, predictions

of operational conditions under which an acceptable sterility assurance level is attained. The disadvantage with this
approach is that it cannot be used in production facilities
because of the risks of microbiological contamination on,
and within, the filling machine and in the manufacturing
area.

The second approach (Ljungqvist and Reinmüller [22, 24]
and Ljungqvist et al. [23])—the "Particle Challenge Test"—is
indirect. It uses a technique that increases the particle level
in ambient air around the critical region or equipment, and
at the same time, measures the particle concentration in
the critical region with a particle counter. The advantage of
this approach is the uncomplicated, immediate registration
of results. As the critical regions only become contaminated
with particles, and not with microorganisms, this approach
is possible to use in clean pharmaceutical production
facilities.

The fact that the airborne microbial contaminants are
transported on particles means that the same conditions of
airborne dispersion are valid for particles with and without
microorganisms. This implies that, if microbiologically
contaminated particles are emitted in the region of a vortex,
an accumulation can occur.

Knowledge about the interaction between air movements
and the dispersion of contaminants plays a vital role in the
microbiological assessment of clean zones protected by
unidirectional air flow. The smoke photography technique
provides valuable information, and combined with the
results from the Particle Challenge Test, it provides a fast
and reliable picture of the potential contamination hazards.
The possibility of entrainment of ambient air into critical
areas or emission of particles (e.g., from operators' activity)

with the risks of accumulation can be visualized, measured, and evaluated.

This approach is performed in three steps. The first step is to visualize the main air movements and identify critical vortex or turbulent regions by using the smoke technique. The second step is to place the probe for particle counting in the critical area and, during measuring, generate particles in the ambient air, e.g., by using Air Current Test Tubes to a level of more than 300,000 particles equal to or larger than 0.5 µm per cubic foot (challenge level). These measurements should be carried out during simulated production activity with operating equipment and personnel interfering with critical movements that are necessary for the process. This simulation should, to some extent, exaggerate the human interference during the measuring periods. The third step is to calculate the Risk Factor.

By defining the ratio between particle concentration in the critical region and particle concentration in ambient air during the Challenge Test, the Risk Factor is obtained. When the Risk Factor is less than 10^{-4} (0.01%), there should be no microbiological contamination from the air in the process according to the authors' experience. If the Risk Factor is higher, remedial, corrective, and preventive actions should be taken, such as change of equipment design, change of working procedures, and so on, before new confirming tests are carried out.

The concepts of Visualisation of Air Movements, Particle Challenge Test, and calculation of the Risk Factor present a method with Limitation of Risks and is known as the LR-Method. This method has been used during Installation Qualification (IQ) as well as during Performance Qualification (PQ). During the IQ, the LR-method was used to discover entrainment of ambient air into a clean zone through openings between side walls and working surface due to activities in the room. During the PQ the LR-method was used to establish manual operations and interference regions allowed within the clean zone area during production.

The LR-Method has been used successfully by the authors for microbiological assessment in clean rooms, especially in critical regions during aseptic processing of sterile drug products. Several different, independent and complicated aseptic processes have been investigated and have had Risk Factors calculated. The critical regions had HEPA-filtered unidirectional class 100 air. The ambient class 10,000 clean room had HEPA-filtered air and ordinary mixing ventilation. Experience from these cases has shown that, where the Risk Factor has been less than 10^{-4}, the final validation with media fills has shown acceptable results.

5.2 Prestudies of Safety Cabinets (Class II)

Three cases with processes sensitive to contamination, performed in microbiological safety cabinets (Class II) that have been checked with the LR-Method, will be described in this section. The principle of the safety cabinet is shown in Figure 22.

The safety cabinets under investigation were checked during installation. They fulfilled the requirements for safety cabinets. Smoke visualization in empty cabinets showed that the main air movements were similar in all three cases.

During the Particle Challenge Test, the measuring probe was placed 15 cm from the front and 10–30 cm over the working surface, depending on where the critical region for the process was determined to be. The front opening was set

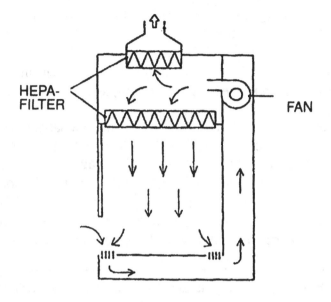

Figure 22. Principle of a safety cabinet.

between 20 and 25 cm, which is the normal aperture during working conditions. Particles, equal to or larger than 0.5 μm, were generated by Air Current Test Tubes in order to reach a concentration of not less than 300,000 particles per ft^3 in the ambient air.

The production cases are identified as A, B, and C, where

A   Consists of a process with large equipment and occasional work with a gas burner. The work station is situated in a class 100,000 region.

B   Consists of a process with large equipment similar to case A but without a gas burner. The work station is situated in a class 100,000 region.

C   Consists of a process with small equipment without a gas burner. The work station is situated in a class 10,000 region.

Results from these measurements are shown in Table 1. Maximum particle levels from samples of 1 ft^3 are reported.

Table 1 shows that cases B and C have a satisfactory Risk Factor, while case A has a Risk Factor of 1.5×10^{-2}, which indicates a potential microbiological hazard. The frequency of microbiologically contaminated products was not satisfactory in case A. Microbiological studies of the bioburden inside and outside the safety cabinet did not, as in the case of the Particle Challenge Test, detect the difference in risk

Table 1. Particle Levels and Calculation of Risk Factor During Installation Qualification (IQ) and Performance Qualification (PQ)

Case	Condition	Measured particle level: No. of particles ≥0.5 μm/ft^3 within the bench	Risk Factor
A	Empty cabinet, IQ	<30	$<1 \times 10^{-4}$
	Simulated activity and disturbances, PQ	400–4500	1.3×10^{-3} 1.5×10^{-2}
B	Empty cabinet, IQ	<10	$<3 \times 10^{-5}$
	Simulated activity and disturbances, PQ	10–30	$(0.3–1) \times 10^{-4}$
C	Empty cabinet, IQ	<10	$<3 \times 10^{-5}$
	Simulated activity and disturbances, PQ	<10	$<3 \times 10^{-5}$

No fewer than 300,000 particles ≥0.5 μm/ft^3 in ambient air during the Challenge Test.

between the various situations. Further studies of the sources of the potential hazards were performed. The results are shown in Table 2.

Weak links in the process could thus be identified, evaluated, and corrected. The effect of remedial actions could be followed and evaluated by the direct method of the Particle Challenge Test.

5.3 Prestudies of Aseptic Filling Lines

Clean room requirements are demanding when stringent particulate control and aseptic conditions are required (see, e.g., Leary [25] and Lhoest [26]). This applies in particular to filling lines located in HEPA-filtered unidirectional air flow. In such cases, wake regions that cause accumulation of contaminants can play critical roles. For example, an elevated stopper bowl situated above or close to critical regions on

Table 2. Mapping of Sources of Potential Risks

Condition	Risk Factor Performance Qualification
Burner on and simulated activity	$(1.3–2) \times 10^{-3}$
Burner on and opening of door	1×10^{-2}
Burner on and rapid passage behind operator	1.5×10^{-2}

Case A during prestudies of Performance Qualification and calculated Risk Factor. No fewer than 300,000 particles ≥ 0.5 μm/ft^3 in ambient air during the Challenge Test.

the filling line can easily destroy all good benefit of a properly designed filling machine. On the other hand, a filling machine designed with no or little regard to aerodynamic demands can cause enormous difficulties from a microbiological safety point of view.

These microbiological problems normally depend on the interaction between air movements and dispersion of contaminants. Examples of in-feeding tables, stopper bowls, and out-feeding stations will be discussed. According to the experience of the authors, rotating feeding tables, due to their large horizontal table surface, cause wake regions that sometimes cause entrainment of ambient air into clean zones. Furthermore, the rotating movements of the table surface can increase the risks. These effects can be reduced when straight feeding tables with smaller horizontal surfaces are used. Particle Challenge Tests have been performed on both types of feeding tables. The results from these measurements confirm there is less risk with a straight feeding

Figure 23. Rotating feeding table.

table than with a rotating table. Figures 23 and 24 show a rotating feeding table and a small straight feeding table, respectively.

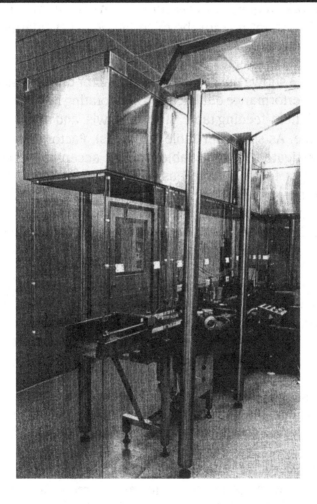

Figure 24. Small straight feeding table.

If the process requires the product vials to be only partly closed when leaving the filler at the out-feeding station, it is absolutely necessary to make certain that microbiological safety is not endangered. Here again, knowledge regarding the wake region and its accumulation of contaminants plays an important role for the microbiological evaluation of the risk situation.

Table 3 shows the calculated Risk Factor during installation and performance qualification of a rotating in-feeding table, straight in-feeding table, stopper bowls, and an out-feeding device. As shown in Table 3, the Risk Factor values of the investigated rotating table are not acceptable, while the straight feeding table has satisfactory values. This is in agreement with the results from media fills and other microbiological tests. The high Risk Factor value below the rotating feeding table depends on the accumulating effect of a large wake vortex.

Different Risk Factor values for stopper bowls are given in Table 3. The differences mainly depend on the shape of the bowl, slanting or straight; on the size of table; and on whether the side of the bowl is perforated to allow air-flow. Each shape has its advantages and disadvantages.

The shape of the bowl, the location in combination with the filling line, the movement of ambient air, and the procedure of filling stoppers into the bowl must be evaluated separately.

The Risk Factor values of the investigated out-feeding device show that there is no risk for entrainment of ambient air (IQ-values), while the allowed interference region is limited to a distance of 15 cm from the device (PQ-values). This limitation depends on a wake region close to the feeding device (see Figures 25 and 26).

Table 3. Calculated Risk Factor During Installation Qualification (IQ) and Performance Qualification (PQ)

Equipment	Location	Risk Factor	
		IQ	PQ
Circular rotating feeding table	above the table below the table	6×10^{-1} 1.6	
Straight feeding table	above the table	$<10^{-4}$	$<10^{-4}$
Stopper bowl	along the edge slanting shape large table	ca 10^{-3}	
	along the edge slanting shape smaller table	$<10^{-4}$	$<10^{-4}$
	along the edge straight shape perforated bowl	$<10^{-4}$	$<10^{-4}$
Out-feeding device	along the edge 5 cm below	$<10^{-5}$	ca 10^{-3}
	5 cm below and 15 cm away	$<10^{-5}$	$<10^{-4}$

No fewer than 300,000 particles ≥ 0.5 $\mu m/ft^3$ in ambient air during the Particle Challenge Test.

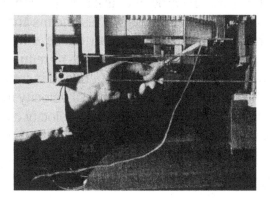

Figure 25. Dispersion of smoke in a unidirectional air flow unit at a distance of 15 cm from and beside an out-feeding device (allowed interference region during operation).

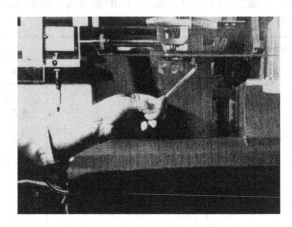

Figure 26. Dispersion of smoke in a unidirectional air flow unit side with and below an out-feeding device (not allowed interference region during operation).

In roboticised production, the movements of the robots are often rapid. When a robot is used in HEPA-filtered unidirectional air flow, the movements can have higher velocities than that of unidirectional air. This means that, in some cases, the robot movements can cause entrainment of ambient room air into the clean zone. To reduce this risk it might, in such cases, be advantageous to increase the velocity of the unidirectional flow up to a velocity in the range of 0.5–1 m/s. The need for increased velocity can be easily demonstrated with the aid of the LR-Method. Air velocity that is finally selected depends on the actual production situation.

5.4 Design of Side Walls

Side walls are used as restricted access barriers around the critical zone and as physical barriers for the protection of operators as well as product. The side walls can be either rigid or flexible. From the standpoint of controlled air movements, rigid walls are always preferred. To allow necessary access, there can be specially designed openings. The length of the side walls should be chosen with regard to necessary product safety, protection from moving parts, necessary access, and correct air movements.

The velocity of the air flow out of the critical zone should be high enough to give protection from ingress of room air during production conditions. The velocity depends on the area of the outlet opening of the side walls. Higher protection will be obtained with higher outflow velocity, i.e., smaller outlet openings. On the other hand, if the openings are too small, stagnation regions will occur in the critical zone. In these regions, there will be uncontrolled air movements with accompanying contamination risks (see Figure 12). Sufficient outlet openings could easily be established with the aid of the LR-Method. By experience, the velocity in the main stream direction through the outlet openings of the side walls should be at least 0.7 m/s.

If the side walls are not directly connected to the HEPA-filters, there will be a certain air flow escaping outside the side walls into the cleanroom. The amount of escaped air depends on the total resistance of air flow/currents over the critical zone inside the side walls. This arrangement will mostly produce less well-defined air movements in the critical zone, and contaminants might not be directly transported out of the critical zone. Due to this, the air movements within and around the critical zone should always be established, and the risk situation should be evaluated with the LR-Method.

As an example, it should be mentioned that, for the production case described in Figure 12, the region that was sensitive to contamination had a critical level 10 cm above the working surface. This means that, when the irrotational vortices reached about 20 cm above the working surface, the production sensitive to contamination was in the vortex region. By studying the vortex structure and the stagnation region at different heights of the side walls and by designing the fixture connecting the two HEPA-filters more aerodynamically, it was possible to place the stagnation region below the critical height of 10 cm and, in this way, ensure safe production.

5.5 HACCP System

HACCP stands for Hazard Analyses Critical Control Point and is an analytical tool that uses a systematic assessment of all steps in the manufacturing operation and the identification of those steps that are critical to the safety of the product. HACCP has become an accepted method for assuring the safety of products. HACCP means

- Conduct Hazard Analyses—identify potential hazards

- Identify the Critical Control Points (CCPs)

- Establish target levels and tolerances

- Establish monitoring systems to ensure control of the CCP by scheduled testing

- Establish corrective actions to be taken when deviations occur

- Establish procedures for verification

- Document all procedures and records

The LR-Method can be used as a tool for evaluating the risk of airborne contamination during the Hazard Analyses (see Ljungqvist and Reinmüller [27]). The whole process chain for the aseptic production of sterile drugs—after sterilisation steps—could be investigated and evaluated step by step with the LR-Method. This could involve the filtration step with its tubing/vessel connections, the transports within the aseptic area, and the loading and unloading of components and products. If the Risk Factor is greater than 10^{-4} during any step, remedial, corrective, and preventive actions should be taken, such as change of equipment design, change of working procedures, and so on, before new confirming tests are carried out.

After the assessment of a potential microbiological risk—i.e., Risk Factor and exposure time—the Critical Control Points (CCPs) should be selected. The LR-Method indicates what, when, and where to monitor the process. The approach with an objective choice of CCPs is to limit the monitoring to those regions where potential hazards occur—areas with high Risk Factors—instead of monitoring a large number of, for example, symmetrically chosen sampling sites. With correctly chosen CCPs, the information will be more significant, and the number of measuring locations can be reduced compared to the common procedures of today.

5.6 Monitoring of Air in Unidirectional Flow Units

When it comes to monitoring the microbiological quality of air in unidirectional flow units, Biotest Diagnostics, Inc.'s (Denville, NJ USA) air sampler, the Reuter Centrifugal Sampler® (RCS), is often used. Tests described by Ljungqvist and Reinmüller [5, 28] in a vertical flow bench with an air velocity of 0.45 m/s clearly show that the critical region (region with turbulence) around the RCS® air sampler is much larger than that of undisturbed parallel air flow.

With a point source (particles) 30 cm above the working surface, an approximate calculation gives a ratio of about 20 between the contaminated area caused by the turbulence of an operating RCS® and an undisturbed parallel flow at the working surface.

The recently developed air sampler from Biotest, the RCS Plus®, does not cause the described type of turbulence from the air intake. However, the outgoing air leaves the RCS Plus® in the form of two jets in opposite directions. As the air in the jets has a higher velocity than the 0.45 m/s common in unidirectional flow units, the jets disturb the parallel flow of clean air, which means that the location of the air sampler during production is still of importance for both process and product.

The schematic views of air patterns of RCS® and RCS Plus® in a vertical unidirectional air flow with a velocity 0.45 m/s are shown in Figure 27. Measurements show that when the operating RCS® air sampler was placed close to a side wall in a vertical flow unit with a working aperture of 30 cm particles from ambient air were transported into the clean zone, though the RCS Plus® did not indicate this type of

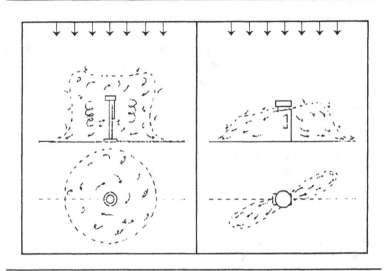

Figure 27. Schematic views of air patterns of RCS® (left) and RCS Plus® (right) in a vertical unidirectional air flow with a velocity of 0.45 m/s.

entrainment. Results from Particle Challenge Tests and the calculated Risk Factors are shown in Table 4.

In conclusion, it must be mentioned that other types of air samplers can cause similar problems. It is important to know about the disturbances caused by air samplers when active air sampling is performed in critical regions. For example, the location of an air sampler ought to be evaluated as well as other types of interference or disturbance. The placing or removing of an air sampler, starting/stopping the sampler, or the turbulence caused by the sampler itself in a unidirectional flow during aseptic operations should be validated in a suitable way. The need for validation applies to all types of air sampling devices when sampling is carried out within a region sensitive to contamination and supplied with unidirectional air flow.

Table 4. Measured Particle Levels and Calculated Risk Factor

Condition	Measured particle levels: No. of particles >0.5 μm/ft³		Risk Factor
	Within the bench	In ambient air	
Empty bench.	<10	>300,000	$<3 \times 10^{-5}$
RCS® air sampler located close to the front aperture. Sampler not operating.	<10	>300,000	$<3 \times 10^{-5}$
RCS® air sampler located close to the front aperture. Sampler operating.	600–30,000	>300,000	$2 \times 10^{-3}–1 \times 10^{-1}$
RCS Plus® air sampler located close to the front aperature. Sampler not operating.	<10	>300,000	$<3 \times 10^{-5}$
RCS Plus® air sampler located close to the front aperature. Sampler operating.	<10	>300,000	$<3 \times 10^{-5}$

Measured during active sampling in a vertical unidirectional air flow unit, when RCS® and RCS Plus® air samplers are located close to a working aperture.

6

WEIGHING STATIONS

6.1 Horizontal and Vertical Air Flow

In work places where pollutants are generated and where it is not possible to use fume cupboards or extraction hoods, ventilation is often arranged as in Figure 28. Extraction of the contaminated air takes place over a relatively large area that is situated near the work station. Distribution of air over the surface should be as uniform as possible. In most cases, air is supplied to the room through ceiling diffusers, through a perforated ceiling, or through a large distribution surface placed adjacent to the opposite wall. The following example illustrates a possible situation at such work places.

At a pharmaceutical industry weighing station, ventilation has been provided according to Figure 28, air inlet by Alt. II. Figure 29 illustrates a situation where dispersion of the pharmaceutical substances has been photographed using special slit lighting.

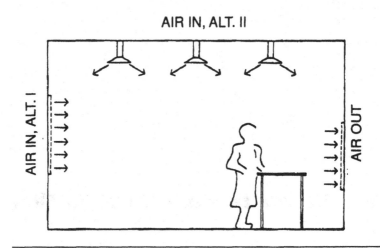

Figure 28. Usual arrangement at a workplace for hazardous work.

Figure 29 shows quite clearly that the presence of a person creates a wake region accompanied by a vortex formation that might extend into the respiratory zone of the person. The wake formation process illustrated here has spoiled the intended beneficial effect of the ventilation system. Investments made with the best intentions have, in this case, been wasted.

Good protection can be obtained by means of a well ventilated work station. This is characterized by creating a vertical principal flow field with extraction through perforated plates in a U-shaped arrangement that is partly mobile. To ensure clean air and well-defined air movements, the weighing station should be equipped with a supply of HEPA-filtered air from the ceiling. Figures 30 and 31 show the dispersion of smoke, both with and without an operator at work.

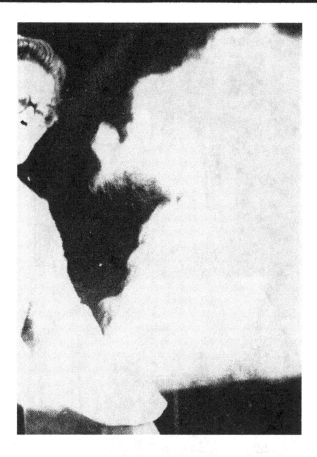

Figure 29. The situation at a weighing station for pharmaceutical substances. The preparation has been photographed by using slit lighting.

The photographs in Figures 30 and 31 illustrate that the air movements are well defined and that the smoke during weighing directly reaches the exhaust air device. Tests have shown that the air velocity in the vertical flow field should be greater than 0.35 m/s to guarantee adequate safety. This

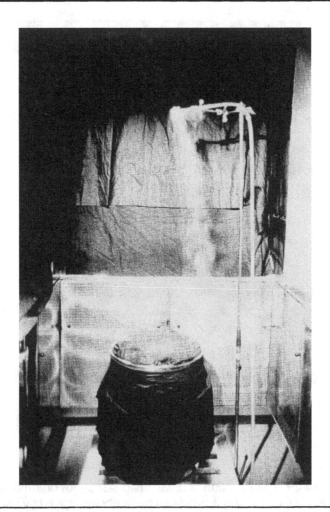

Figure 30. Dispersion of smoke in a weighing station without an operator.

is in agreement with the values in the region of 0.3–0.4 m/s given by Whyte et al. [29], which are evaluated from measurements of unidirectional air flow systems for orthopedic surgery.

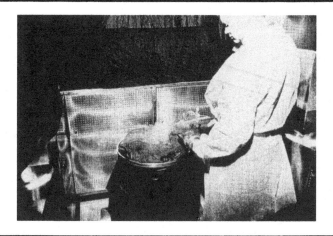

Figure 31. Dispersion of smoke in a weighing station with an operator working.

The exhaust system is constructed with perforated surfaces, where different types of perforation give various pressure differences, e.g., varying exhaust air volume flows per unit surface.

Two types of newly designed weighing stations will be discussed (see Ljungqvist and Reinmüller [30]). The first station is for weighing substances up to 15 kg, and the second station is for weighing portions between 15 and 60 kg. The principal exhaust systems of the two types of weighing stations are shown in Figures 32 and 33.

The total air volume flow through the exhaust system shown in Figure 32 is 3500 m³/h, with about 30% through the perforated horizontal surfaces, 15% through the vertical back part, and the remaining 55% through the vertical perforated surfaces in the middle, around the drum region. The total exhaust air volume flow through the station shown in Figure 33 is 4500 m³/h, where the air flows through the perforation

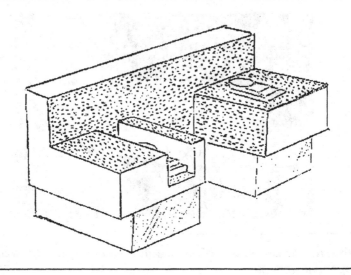

Figure 32. Weighing station equipped with electronic scales (up to 2 kg) or bench scales (2–15 kg).

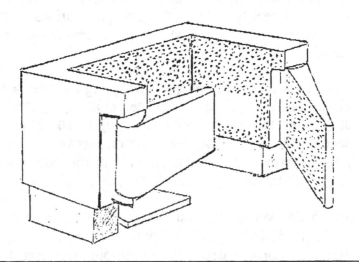

Figure 33. Weighing station for floor located scales (15–60 kg).

in the back part, side parts, and mobile front gates are one-third each.

The supply air volume flow through the HEPA-filters above the exhaust units is about 10% more than that of the exhaust air. Figure 34 shows an example of a principal arrangement with a supply air unit over the exhaust unit shown in Figure 33.

6.2 Qualification with Tracer Gas

In the tracer gas method, nitrous oxide (N_2O) is used as a tracer. The gas is emitted at a constant flow rate (0.13 m^3/h) through a specially designed line source producing an

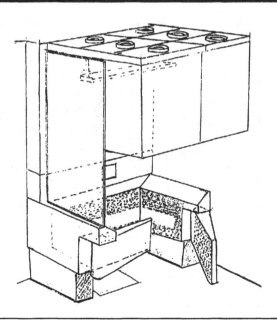

Figure 34. Principal arrangement of a weighing station with supply air (HEPA-filtered) system and exhaust air unit.

almost momentum-free outflow of gas into the working regions of the two weighing stations. More precisely, the tracer gas outlet was placed above the substance drum in the weighing area. It can be placed, alternatively, above various sized drums.

Concentrations of nitrous oxide are measured by using an infrared analyzer (URAS, 3G) at several points and at two levels, 1.0 m and 1.4 m, over the floor and in a vertical plane 0.01 m from the edge of the exhaust device of the weighing stations. Figures 35 and 36 show the positions of the test equipment during tracer gas measuring at the two weighing stations.

Effects of disturbances at the weighing stations can be determined when measurements are made with an operator standing still in front of the working stations and also with

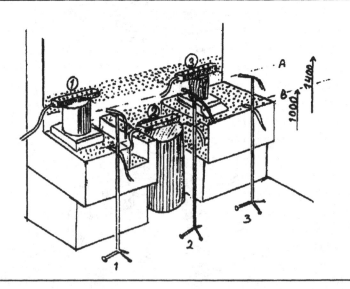

Figure 35. Positions of the test equipment during the tracer gas measuring at the station; used for weighing substances up to 15 kg.

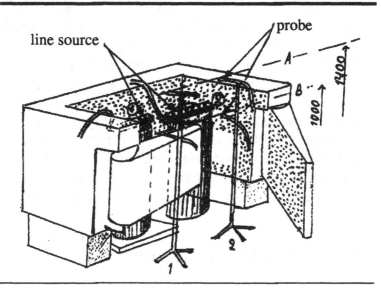

Figure 36. Positions of the test equipment during the tracer gas measuring at the station; used for weighing substances from 15 to 60 kg.

the operator moving his or her hands in a "calm" manner within the working region of the stations. Since the tests were made to evaluate the safety of the weighing station the maximum values have always been used.

The Escape Safety Value, E-value, described by Ljungqvist and Malmström [31] as follows:

$$E = \frac{c_e}{c_e + C_{max}} \tag{10}$$

where c_e = tracer gas concentration in exhaust air (mean value)

        C_{max}= maximum recorded concentration just outside the ventilated station.

The tracer gas method, as it is used here, takes into consideration the disturbances caused by an operator at work. The violent pulsating concentrations compared to the response speed of the measuring equipment and how the operator moves must be taken into consideration in the evaluation of E-values (Ljungqvist and Malmström [31] and Ljungqvist and Waering [32]). Measurements with the URAS 3 infrared analyzer (MANNESMANN HARTMANN & BRAUN AG, Frankfurt, Germany) from a large number of work places with local exhaust ventilation have, from experience, shown that E-values higher than 95% constitute safe working places.

6.3 Results and Discussion

Measurements without an operator at the two types of weighing stations have not yielded any significant values. In the following, the only values that are discussed are when the operator is standing in front of the stations, either standing still or working in a calm manner. The Escape Safety Values (E-values) measured with the URAS 3 infrared analyzer are shown in Tables 5 and 6.

The results in Tables 5 and 6 show that most E-values are higher than 95%, which constitute safe work places from a personnel protection point of view. In Table 5, when the line source/probe is in position 1, the E-value for an active operator is 85%. This mainly depends on the fact that the scales used during measuring were larger than the scales originally designated to be used in the station. This means that the scales should always be placed at the lower specially designed position. Table 6 illustrates that the E-value becomes lower than 95% when one of the gates is open in the station for weighing substances from 15 to 60 kg.

Table 5. *E*-values in Percent of the Weighing Station for Substances up to 15 kg

Position of line source/probe	E-VALUES (in percent)			
	Operator at rest		Operator moving	
	Height A (1400 mm)	Height B (1000 mm)	Height A (1400 mm)	Height B (1000 mm)
Line source/probe in position 1 (left side, 2–15 kg)	100	100	96.6	85.0
Line source/probe in position 2 (in the middle)	100	100	100	100
Line source/probe in position 3 (right side, 0–2 kg)	100	100	100	98.8

Position of test equipment according to Figure 35.

The LR-Method has been used to evaluate the weighing station for substances used for sterile products. Particle Challenge Tests and calculation of Risk Factors have been performed to assess the microbiological risks to which the product can be exposed.

Weighing of sterile products was performed in a clean room with HEPA-filtered air and ordinary mixing ventilation—class 10,000. The critical regions over the weighing stations have HEPA-filtered air—class 100. The particle counter probe was located in the critical area over the substance drum, and particles were generated close to the working operator.

Table 6. *E*-values in Percent of the Weighing Station for Substances between 15 and 60 kg.

Position of line source/probe	E-VALUES (in percent)			
	Operator at rest		Operator moving	
	Height A (1400 mm)	Height B (1000 mm)	Height A (1400 mm)	Height B (1000 mm)
Line source/probe in position 1 (left side)	100	100	97.2	95.8
Line source/probe in position 2 (right side) gates closed	100	100	100	98.6
Right gate open	100	100	100	87.5

Position of test equipment according to Figure 36.

The results from these measurements show that the Risk Factor is less than 10^{-4} (0.01%) whether the operator is at rest or is performing weighing movements in a calm manner. When the operator is moving in a more boisterous manner the Risk Factor increases. This clearly shows the importance of correct working discipline from the product safety point of view.

Work at weighing stations for pharmaceutical substances can cause risk situations for the operator as well for the product. When operators are working at stations, their movements play a vital role.

The safety of a weighing station cannot be assessed when the stations are evaluated solely on the basis of air flow or air velocity, as these values do not relate to the operators and their movements. The aerodynamic flow pattern into and around the whole weighing station during activity play an important role in any contamination situation, i.e., operator safety and product safety.

The results of the qualification for the two types of weighing stations discussed here show that it is possible to obtain good protection (safety) for the operator as well as for the product. The results also show the importance of good working performance and discipline.

7

AIR FLOW
THROUGH OPENINGS

Air flow through doorways is discussed in several papers, see, e.g., Shaw and Whyte [33], Linden and Simpson [34], Kiel and Wilson [35], and Whyte [36]. The driving mechanisms are often a combination of density differences, mechanical ventilation, motion of a person through the opening, and the motion of the door itself.

In most practical situations, the density differences are caused by temperature differences. In clean rooms the mechanical ventilation must maintain a positive pressure differential (ca 15 Pa) to adjacent less clean areas, when all doors are closed. When a door is open and there is no temperature difference, outward flow generally is sufficient to minimise ingress of contamination. According to Shaw and Whyte [33], the amount of air required to pressurize an area to prevent air passing an open doorway would be extremely large if the temperature differences were not kept small. A flow of about 0.25 m^3/s for each square metre of doorway is required for a temperature difference of 1°C.

Experimental results given by Kiel and Wilson [35] show for typical door swing speeds that the pumping exchange could be neglected entirely above a temperature difference of 3–5°C. At a temperature difference of zero, the volume pumped increased linearly with the speed of the moving door, with a typical exchange volume of about 50% of the swept volume of the door.

When small temperature differences occur, the air flow through a doorway can be estimated only approximately from the relationship describing density driven flow. At higher temperature differences (>4°C) the estimation will be more correct. The velocity profile through a doorway with temperature difference is schematically shown in Figure 37.

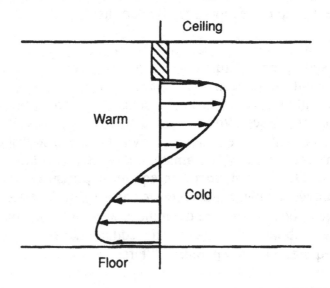

Figure 37. Schematic representation of the velocity profile through a doorway with temperature difference.

Through one-half of the opening the total discharge flow rate (in both directions) will be

$$Q = C_d \frac{WH^{3/2}}{3} \left(g \frac{\Delta \rho_0}{\rho_{0m}}\right)^{1/2} \tag{11}$$

where C_d = discharge coefficient

$\quad\quad\;\; W$ = opening width

$\quad\quad\;\; H$ = opening height

$\quad\quad\;\; g$ = gravitational acceleration

$\quad\quad\;\; \Delta\rho_0$ = density difference

$\quad\quad\;\; \rho_{0m}$ = mean density

The coefficient C_d has experimentally been estimated to 0.8 by Shaw and Whyte [33]. A mixing region between the two air flows, see Figure 37, will occur if the interface is unstable, according to Kiel and Wilson [35], and the discharge coefficient will be replaced by an overall orifice coefficient with a value about 0.6. In the following, the coefficient 0.8 is chosen, because the higher air flows indicates higher contamination risks.

With the aid of the equation of state for an ideal gas, the density relation in Equation (11) can be expressed as a function of temperature:

$$\frac{\Delta\rho_0}{\rho_{0m}} = \frac{2\Delta T}{(T_1 + T_0)} \tag{12}$$

where ΔT = temperature difference

T_1 = temperature

T_0 = reference temperature

Graphical representations of Equation (11) in combination with Equation (12) expressed in flow rate as a function of temperature difference and opening dimensions are shown in Figures 38–40. The reference temperature (0 in diagrams) is chosen to normal room temperature 20°C (293K). Figure 39 should be used for increasing temperature and Figure 40 for decreasing temperature. Figure 38 can, in practical matters, be used for increasing as well as decreasing temperature.

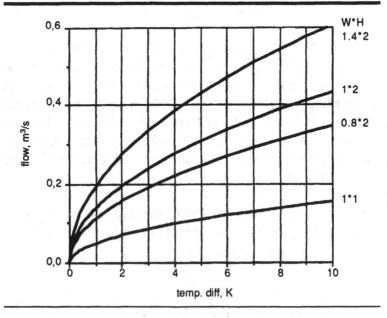

Figure 38. Flow rate as a function of temperature difference (ref. temp. 20°C) and opening dimensions (width *W* and height *H* in m).

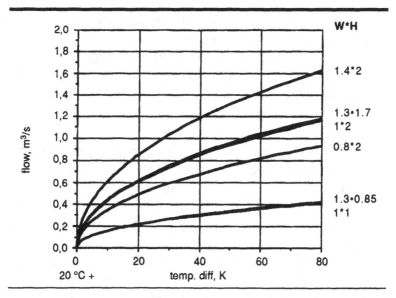

Figure 39. Flow rate as a function of temperature difference (increasing temp with ref. temp. 20°C) and opening dimensions (width *W* and height *H* in m).

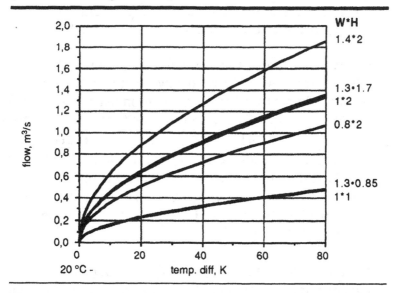

Figure 40. Flow rate as a function of temperature difference (decreasing temp. with ref. temp. 20°C) and opening dimensions (width *W* and height *H* in m).

Figure 38 shows that for ordinary door openings, even at small temperature differences of 2–5°C, the discharged flow rate will be relatively high, 0.15–0.3 m³/s (500–1000 m³/h). This should be taken into consideration when contamination risks are discussed in clean rooms with temperatures different from adjacent less clean areas.

In pharmaceutical manufacturing some processes often cause temperature differences in the surrounding air, e.g., sterilization in autoclaves and freeze-drying processes. Generally there is a temperature difference when the door of the unit is opened, and this causes flow of room air through the opening, and a contamination risk can occur. To avoid this risk, a HEPA-filter unit should be installed above the opening to provide clean air close to the opening. The air flow needed through the HEPA-filter depends on the temperature difference, and the flow of clean air should be greater than that of the theoretical calculated flow. If there are side walls or restrictions in the operators' movements, the risk could be estimated with aid of the LR-Method.

For example, a commonly used freeze-dryer with an opening 1.3 m wide and 1.7 m high can have a temperature difference of 60°C relative to the clean room. From Figure 40 (decreasing temperature), it can be estimated that the air flow is about 1.15 m³/s (4140 m³/h). It is recommended that the HEPA-filter unit above the opening of the freeze-dryer should, during the time the freeze-dryer is open, be able to give an air volume flow at least 20% higher than that of theoretical estimated, i.e., in this case 1.38 m³/s (ca. 5000 m³/h).

If instead the freeze-dryer has double doors, each 1.3 m × 0.85 m, and only one door at a time is opened, the theoretical air volume flow in Figure 40 can be estimated to 0.4 m³/s (1440 m³/h). The installed HEPA-filter unit should at least be able to give an air volume flow of 0.48 m³/s (ca 1700 m³/h).

The air velocity measured just below the HEPA-filter is normally 0.45 m/s. This gives a filter area of about 3.1 m² for the case with the whole door and about 1.1 m² for the case with double doors. In the first case, it should be considered whether a fan with two speeds should be chosen. The higher air flow should only be used during the initial, critical opening situation, and if that higher air velocity is acceptable, a smaller HEPA-filter unit could be sufficient.

If a HEPA-filter unit will be installed above an autoclave, Figure 39 with increasing temperature, should be used. For example, an autoclave with door dimensions of 1 m × 1 m and 30°C higher inside temperature than that of the clean room requires—during the opening of the autoclave—a theoretical air flow of about 0.25 m³/s (900 m³/h). Here again the HEPA-filter unit used should, during the time the autoclave door is opened, give an air flow 20% higher than that of the estimated air flow, i.e., 0.3 m³/s (ca. 1100 m³/h).

Finally it should be reiterated that, as a part of the Installation Qualification and Performance Qualification, the LR-Method should be used to estimate whether the installations will ensure safe production. Questions such as the length of the side walls around the HEPA-filter unit and allowed operator movements during loading and unloading can easily be answered by using the LR-Method.

8

CONCLUSION

When unidirectional air flow benches/units are used for processes sensitive to contamination, a thorough function check with equipment in place should always be carried out during the Installation Qualification. Studies of air movements, depicted with visual illustrative methods, give both prompt and valuable information.

In order to reduce the influence of unfavorable stagnation regions and vortex structures with the risk for accumulation of contaminants, visualization tests should be carried out when designing the side walls. In connection with these tests, Particle Challenge Tests should also be performed. Here, particles (smoke) outside the unit and a probe of a particle counter, placed inside the unit for measuring the particle concentration in the critical regions and calculation of the Risk Factor, can give valuable information. The Particle Challenge Test indicates entrainment of ambient air into the clean zone, and in this way, draws attention to the potential microbiological hazards. This entrainment of room air into the clean zone is very difficult to detect with common microbiological test methods.

During activity and as a part of Performance Qualification, investigations should be carried out with the operators performing movements that are necessary for the process. This is to establish manual operations allowed outside the clean zone and to establish interference regions allowed inside the clean zone.

The systematic approach with Visualization of Air Movements, Particle Challenge Tests, and calculation of the Risk Factor presents a method—the LR-Method—to evaluate the risks of human interference in a critical zone, e.g., an aseptic filling line. It also gives valuable information concerning potential weak links.

The LR-Method discussed here is an engineering tool that provides valuable information concerning weak links. The information is of vital importance for hazard analyses and for establishing Critical Control Points. The method also presents results for educational purposes and motivates the operators to act and work in a correct manner. Use of the LR-Method does not replace the final evaluation of aseptic processes with media fills.

In clean rooms with low microbiological burden, the monitoring of viable particles gives results with high relative variation, and the results are often less than one CFU per sample. This means that particle counters for viable and nonviable particles give more statistically acceptable results than that of air samplers for viable particles. Furthermore, the particle counter systems could often be used for continuous monitoring measurements, yielding valuable information about different activities in clean rooms. Deviation from normal levels of airborne contaminations could be detected earlier by particle monitoring than by microbial monitoring.

When a systematic hazard analysis is performed and Critical Control Points are established in a rational manner, today's comprehensive monitoring by microbiological sampling could be improved. The goal of the combination of continuous particle monitoring and microbiological monitoring is to provide maximised information regarding product quality and patient safety. A combination of continuous monitoring and microbiological monitoring gives more reliable information regarding product safety. This approach also reduces the potential risks associated with performing microbiological monitoring in the clean rooms or the clean zones.

In conclusion, the LR-Method gives valuable information regarding the probability of contamination, but does not take into account the exposure time. For example open bottles on a feeding table or stoppers in a stopper bowl with long exposure times can, even if the probability of contamination is low, have the same or higher total risk than, e.g., a filling station, with a short exposure time and higher probability of contamination.

The tracer gas method can be used to evaluate and compare the containment performance of ventilated work stations without recirculation of air. Containment is perceived as being closely related to internal geometry and aerodynamics, as well as the air flow pattern in the work station. The method also gives information regarding operator safety and the importance of good working performance and discipline.

REFERENCES

1. Bird, R. B., Stewart, W. E., and Lightfoot, E. N. *Transport Phenomena*, John Wiley & Sons, Inc., New York, 1960.

2. Fuchs, N. A. *The Mechanics of Aerosols*, Pergamon Press, Oxford, 1964.

3. Hinze, J. O. *Turbulence*, McGraw-Hill, Inc., New York, 1975.

4. Ljungqvist, B. "Some observations on the interaction between air movements and the dispersion of pollution." Document D8:1979, Swedish Council for Building Research, Stockholm, 1979.

5. Ljungqvist, B. and Reinmüller, B. "Some Aspects on the Use of the Biotest RCS Air Sampler in Unidirectional Air Flow Testing," *J. Parenter. Sci. Technol.*, Vol. 45, pp. 177–180, No. 4, 1991.

6. Friedlander, S. K. *Smoke, Dust and Haze, Fundamentals of Aerosol Behaviour*, John Wiley & Sons, Inc., New York, 1977.

7. Hinds, W. C. *Aerosol Technology*, John Wiley & Sons, Inc., New York, 1982.

8. Whyte, W. "Sterility Assurance and Models for Assessing Airborne Bacterial Contamination," *J. Parenter. Sci. Technol.*, Vol. 40, pp. 188–197, No. 5, 1986.

9. Noble, W. C., Lidwell, O. M., and Kingston, D. "The size distribution of airborne particles carrying micro-organisms," *J. Hyg.*, 66: 385, 1963.

10. Clark, R. P., Reed, P. J., Seal, D. V., and Stephenson, M. L. "Ventilation conditions and air-borne bacteria and particles in operating theatres; proposed safe economies," *J. Hyg., Camb.* 95, pp. 325–335, 1985.

11. Ljungqvist, B. "Air Movements—The Dispersion of Pollutants: Studies with Visual Illustrative Methods," *ASHRAE Transactions*, Vol. 93, Part 1, pp. 1304–1317, 1987.

12. Kim, T., and Flynn, M. "Airflow Pattern Around a Worker in a Uniform Freestream," *Am. Ind. Hyg. Assoc. J.*, Vol. 52, pp. 287–296, 1991.

13 Clark, R. P., and Edholm, O. G. *Man and His Thermal Environment*, Edward Arnold Ltd., London, 1985.

14. Busnaina, A. A. "Modeling of Clean Rooms on the IBM Personal Computer," Proceedings, Institute of Environmental Sciences, 33rd Annual Technical Meeting, San Jose, CA, May 4–8, 1987.

15. Busnaina, A., and Ljungqvist, B. "Prediction and Flow Visualization of Air Flow in Clean Environments," *Computers in Engineering*, Vol. 1, Proceedings of the ASME Conference, Santa Clara, CA, August 18–22, 1991.

16. Hiemenz, K. "Die Grenzschicht an einem in den gleichförmigen Flüssigkeitsstrom eingetauchten geraden Kreiszylinder; Thesis Göttingen 1911," *Dingl. Polytech. J.*, 326, 321, 1911.

17. Ljungqvist, B., Nydahl, R., and Reinmüller, B. "Some Observations on Air Movements in Open Unidirectional Air Flow Benches," *Swiss Contamination Control*, No. 4a, pp. 36–39, 1990.

18. Batchelor, G. K. *An Introduction to Fluid Dynamics*, Cambridge University Press, 1970.

19. Schlichting, H. *Boundary-Layer Theory*, McGraw-Hill, Inc., New York, 1979.

20. Ljungqvist, B. and Reinmüller, B. "Some Observations on Environmental Monitoring of Cleanrooms," *European Journal of Parenteral Sciences*, Vol. 1, No. 1, pp. 9–13, 1996.

21. Bradley, A., Probert, S. P., Sinclair, C. S., and Tallentire, A. "Airborne Microbial Challenges of Blow/Fill/Seal Equipment; A Case Study," *J. of Parenter. Sci. and Technol.*, Vol. 45, No. 4, pp. 187–192, 1991.

22. Ljungqvist, B. and Reinmüller, B. "Supplementary microbiological assessment with the aid of particle challenge test and visualization technique," *Proceedings of the 11th International Symposium on Contamination Control*, ICCCS, London, 21–25 September, 1992.

23. Ljungqvist, B., Reinmüller, B., and Nydahl, R. "Prestudies for Microbiological Assessment in Clean Zones," *Proceedings of the 2nd International PDA-Congress*, Basel, Switzerland, 22–24 February, 1993.

24. Ljungqvist, B. and Reinmüller, B. "Interaction Between Air Movements and the Dispersion of Contaminants: Clean Zones with Unidirectional Air Flow," *J. of Parenter. Sci. and Technol.*, Vol. 47, No. 2, pp. 60–69, 1993.

25. Leary, H. R. "Designing Automated High Speed Packaging Lines for Cleanroom Operations," *Pharmaceutical Engineering*, Vol. 12, No. 4, pp. 9–15, 1992.

26. Lhoest, W. J. "Design and Selection of Pharmaceutical Production Equipment in the Scope of Modern Automated Plants—Part I," *Pharmaceutical Engineering*, Vol. 11, No. 2, pp. 46–58, 1991.

27. Ljungqvist, B. and Reinmüller, B. "Hazard Analyses of Airborne Contamination in Clean Rooms—Application of a Method for Limitation of Risks," *PDA Journal of Pharmaceutical Science and Technology*, Vol. 49, No. 5, pp. 239–243, 1995.

28. Ljungqvist, B. and Reinmüller, B. "The Biotest RCS Air Samplers in Unidirectional Flow," *Journal of Pharmaceutical Science and Technology*, Vol. 48, No. 1, pp. 41–44, 1994.

29. Whyte, W., Shaw, B. H., and Barnes, R. "A bacteriological evaluation of laminar-flow systems for orthopaedic surgery." *J. Hyg, Camb.*, 71, pp. 559–564, 1973.

30. Ljungqvist, B. and Reinmüller, B. "Qualification of Weighing Station for Pharmaceutical Substances," *PDA Journal of Pharmaceutical Science and Technology*, Vol. 49, No. 2, pp. 93–98, 1995.

31. Ljungqvist, B. and Malmström, T-G. "Tests of laboratory fume hoods." In *Ventilation 85* (H D Goodfellow, ed.) pp 755–762. Elsevier Science Publishers, B V, Amsterdam, 1986.

32. Ljungqvist, B. and Waering, C. "Some observations on modern design of fume cupboards," In *Ventilation 88* (J. H. Vincent, ed,), pp 83–88. Pergamon Press, Oxford, 1989.

33. Shaw, B. H. and Whyte, W. "Air movement through doorways—the influence of temperature and its control by forced airflow," *Building Services Engineering*, Vol. 42, pp. 210–218, Dec 1974.

34. Linden, P. F. and Simpon, J. E. "Buoyance Driven Flow Through an Open Door," *Air Infiltration Review*, Vol. 6, No. 4, pp. 4–5, 1985.

35. Kiel, D. E. and Wilson, D. J. "Combining door swing pumping with density driven flow," *ASHRAE Transactions*, 95, part 2, 1989.

36. Whyte, W. *Cleanroom Design*, John Wiley & Sons, Ltd. Chichester, 1991.

37.* Sahlin, P., Bring, A., and Sowell, E. F. "The Neutral Model Format for Building Simulation," Bulletin No. 32, Building Services Engineering, Royal Institute of Technology, Stockholm, 1994.

38.* Sahlin, P. and Bring, A. "Applying IDA to Airflow Problems in Buildings," ITM report 1993:2, Swedish Institute of Applied Mathematics, Göteborg, 1993.

39. Isfält, E., Ljungqvist, B., and Reinmüller, B., "Simulation of Airflows and Dispersion of Contaminants Through Doorways in a Suite of Clean Rooms," *European Journal of Parenteral Sciences,* Vol. 1, No. 3, pp. 67–73, 1996.

* the NMF reference report and other articles can be fetched at ftp://urd.ce.kth.se/pub/reports

SYMBOLS

a	Distance between point source and the center of vortex (origin), cm, m
A_e	Area of exposed surface, cm^2, m^2
c	Concentration; particles, number/cm^3, number/m^3, gases, cm^3/cm^3, m^3/m^3 or ppm
c_o	Initial concentration, number/cm^3, number/m^3
c_b	Concentration of bacteria-carrying particles, number/cm^3, number/m^3
c_e	Tracer gas concentration in exhaust air (mean value), ppm
c_m	Mean value of concentration over a closed streamline (circle); particles, number/cm^3, number/m^3, gases, cm^3/cm^3, m^3/m^3 or ppm
c_{max}	Maximum recorded concentration just outside the ventilated station, ppm

C_d Discharge coefficient, nondimensional

D Diffusion coefficient, cm^2/s, m^2/s

E Escape safety value, nondimensional

F Nondimensional factor

F_l Nondimensional factor

g Gravitational acceleration, m/s^2

H Height, cm, m

$\mathbf{K}$ Flux vector; particles, number/(cm^2,s), number/(m^2,s), gases, $cm^3/(cm^2,s)$, $m^3/(m^2,s)$, or ppm cm/s, ppm m/s

K_s Particle flux, monodispersal aerosol settling, number/(cm^2,s), number/(m^2,s)

K_x Particle flux in x-direction, number/(cm^2,s), number/(m^2,s)

K_y Particle flux in y-direction, number/(cm^2,s), number/(m^2,s)

K_z Particle flux in z-direction, number/(cm^2,s), number/(m^2,s)

$K\varphi$ Particle flux in φ-direction, number/(cm^2,s), number/(m^2,s)

K_ρ Particle flux in ρ-direction, number/(cm^2,s), number/(m^2,s)

K_{p1} Particle flux in the direction perpendicular to the stream line, which passes through the point source in a unidirectional air flow, number/(cm^2,s), number/(m^2,s)

N_d Number of bacteria-carrying particles deposited onto a surface, number

N_s Number of particles of a monodisperse aerosol settling, number/cm^2, number/m^2

q Outward flow from a point source; particles, number/s; gases, cm^3/s, m^3/s

Q Flow rate, m^3/s

r Distance to the point source, cm, m $r = (x^2 + y^2 + z^2)^{1/2}$

t Time, s

t_e Exposure time, s

T_0 Reference temperature, K

T_1 Temperature, K

ΔT Temperature difference, K, °C

v Velocity vector, cm/s, m/s

v_o Constant air velocity in the x-direction, cm/s, m/s

v_s Settling velocity, cm/s, m/s

v_x Velocity in x-direction, cm/s, m/s

v_y Velocity in y-direction, cm/s, m/s

v_z Velocity in z-direction, cm/s, m/s

W Width, cm, m

x, y, z Positional coordinates

α Modular angle in the complete elliptic integral of the first kind in Equation (A.15)

ξ Nondimensional positional coordinate

ρ, φ Radius and angle in cylindrical coordinates

ρ_{0m} Mean density, kg/m³

$\Delta\rho_0$ Density difference, kg/m³

τ Nondimensional quantity

Ω Nondimensional quantity

ω Angular velocity, 1/s

GLOSSARY

Aerosol—Liquid or solid particles suspended in a gas, usually air.

Air Changes—The frequency per unit time (minutes, hours, etc.) that the air within a controlled environment is replaced. The air can be recirculated partially or totally replaced.

Air Sampler—Devices or equipment used to sample a measured amount of air within a specified time to determine the particulate or microbiological status of air in the controlled environment.

Aseptic—Literally, "without infectious matter," but often erroneously used as a synonym for "sterile" (i.e., absence of all living organisms).

Aseptic Area—The space immediately surrounding the critical zone or area of an aseptic fill operation. Must conform to at least Class M4.5 contamination level.

Aseptic Filling—Sterile containers are filled with sterile product and sealed with sterile closures. Frequently used with products that cannot tolerate terminal sterilization.

Aseptic Technique—All practices that avoid micro-biological contamination.

Aseptic Processing—A mode of processing pharmaceutical and medical products that involves the separate sterilization of the product and of the package (container/closure or packaging material for medical devices), and the transfer of the product into the container and its closure under microbiological critically controlled conditions.

Autoclave—A device for sterilising items with steam under pressure.

Barrier—Any physical obstacle that separates and protects against contamination.

Centrifugal Air Sampler—An impaction-type microbiological sampler that captures viable particles by accelerating them into a nutrient surface through a centrifugal blower. The RCS® or Reuter sampler is an example of a centrifugal sampler.

Clean Room—A room in which the concentration of airborne particles is controlled to meet a specified airborne particulate cleanliness class. In addition, the concentration of microorganisms in the environment is monitored; each cleanliness class defined is also assigned an acceptable microbiological level for the air, surfaces, and personnel gear.

Clean Zone—A defined space in which the concentration of airborne particles and microorganisms is controlled to meet specific cleanliness class levels.

Cleanliness Class—See Federal Standard 209E.

Colony—A discrete, visible accumulation of microbial growth on the surface of a solid culture medium.

Colony Forming Unit (CFU)—That which results in the formation of a colony of microbial growth on appropriate solid media.

Controlled Environment—Any area in an aseptic process system for which airborne particulate and microorganism levels are controlled to specific levels, appropriate to the activities conducted within that environment.

Critical Area/Region/Zone—Space immediately surrounding an aseptic filling operation.

DOP (Dioctylphtalate)—A chemical capable of producing smoke of a very uniform particle size. Frequently used for testing air filters. Chemical synonym is di(2-ethyl-hexyl)phtalate.

E-Value—Escape Safety Value is defined as the quotient between mean value concentration in the exhaust air and the sum of the mean value concentration in the exhaust air and maximum concentration just outside the ventilated station.

(U.S.) Federal Standard 209E—"Airborne Particulate Cleanliness Classes in Clean Rooms and Clean Zones" is a standard approved by the Commissioner, Federal Supply Services, General Service Administration, for use by "All Federal Agencies." The Standard establishes classes of air cleanliness based on specified concentration of airborne particulates. These classes of air cleanliness have been developed, in general, for the electronic industry's, "super-clean" controlled environments. In the pharmaceutical industry, Federal Standard 209E is used to specify the construction of a controlled environment. Class 100, Class 10,000, and Class 100,000 are generally represented in an aseptic processing system. If the classification system is applied on the basis of particles equal to or greater than 0.5 μm, these classes are now represented in the SI system by Class M3.5, M5.5, and M6.5, respectively.

Filter Integrity—Characteristics of a filter that ensure the functional performance of a filter used for liquid or gas in an aseptic processing system.

Flux Vector—In principle the numerical value of the flux vector indicates the number of particles passing a supposed unit area, which has been placed perpendicular to the direction of particle flow, per unit time.

Freeze-Dryer—A device for freeze-drying; in its simplest form it consists of a vacuum chamber into which wet material can be placed, together with a means of removing water vapor so as to freeze the material by evaporative cooling, and then maintain the water vapor pressure below the triple point pressure.

Freeze Drying—A process that consists of three separate, unique, but interdependent, processes: freezing, sublimation, and desorption.

HACCP—Hazard Analyses Critical Control Point. It is an analytical tool that involves a systematic assessment of all steps in the manufacturing operation and the identification of those steps that are critical to the safety of the product.

HEPA-Filter—High Efficiency Particulate Air filter.

Laminar Flow—A type of flow characterized by a smooth flow, free of any disturbances, such as, small and temporary vortices, eddies.

LR-Method—A method for Limitation of Risks. It is a concept that includes air movements depicted visually, the Particle Challenge Test and calculation of Risk Factor.

Media Fill Test—A validation test for aseptic filling by using sterile growth medium in place of product and looking for microbiological growth in the filled containers following appropriate incubation.

Microbiological Burden—The level of airborne micro-organisms in the air.

Mixing Air—The concept of mixing air in a room, whereby complete mixing is considered to be achieved, is here called "complete mixing of air" or "completely turbulent mixing" or "fully turbulent air."

Monodisperse Aerosol—An aerosol which has particles that are all the same size.

Open Unidirectional Air Flow Bench—Containment device with vertical or horizontal HEPA-filtered unidirectional air for protecting the work area from environmental contamination.

Particle Challenge Test—An indirect test, which implies a technique with an increase of the particle level in ambient air around the critical region, and at the same time to measure the particle concentration in the critical region with a particle counter.

Particle Flux—The number of particles that pass a supposed unit area, which has been placed perpendicular to the direction of particle flow, per unit time.

RCS®—Reuter Centrifugal Sampler (see Centrifugal Air Sampler).

Risk Factor—A factor defined by the ratio between particle concentration in the critical region and the particle concentration in ambient air during the Particle Challenge Test. When the Risk Factor is less than 10^{-4} (0.01%) there should be no microbiological contamination from the air during process conditions (clean room application).

Safety Cabinet (Class II)—Containment device for protecting the work area from environmental contamination and the operator from hazardous microorganisms.

Settling Plate—A method for assessing microbiological fallout by leaving uncovered petri dishes containing growth medium in selected locations for a predetermined period of time. Following appropriate incubation, the numbers of colonies that form on the agar are an indication of how effective the air movement in a bioclean facility is in preventing airborne contamination from settling onto work or product surfaces.

Spore—A subcellular body that some species of bacteria form and that is considerably more resistant to harsh conditions, such as heat, disinfectants, and radiation, than the vegetative form. Bacterial spores are more correctly referred to as endospores to distinguish them from the spores formed by molds, which are somewhat less resistant to harsh conditions.

Sterile—Free of all living organisms.

Sterilization—The complete removal or destruction of all living organisms.

Turbulent Flow—A type of flow characterized by small temporary fluctuations caused by instabilities.

Unidirectional Flow—A uniform parallel flow, which can be either laminar or turbulent. The flows in most practical cases are turbulent.

Validation—Full, detailed documentation that all processes and procedures are functioning in the manner in which they were designed.

Viable—Capable of life. The ability of a microorganism to grow and form visible colonies on an appropriate medium.

Viable Particle—A particle, such as dust, lint, or skin cells, that has one or more viable microbial cells on it.

Vortex—A vortex is characterized by the fact that the streamlines are closed within a region. Two main types exist: 1) Free (irrotational) vortex characterized by the fact that the velocity varies with the radius and increases towards the center. 2) Forced or rotational vortex characterized by the fact that the velocity decreases linearly to the center; also called rigid body rotation.

Vortex Street—A regular pattern of vortices in a parallel flow being shed downstream from alternative sides of an obstacle.

Wake—A mixing zone created by an obstacle in a parallel flow field, characterized by eddies or vortices which entrain air into a reverse-flow near the obstacle.

Appendix 1

MATHEMATICAL TREATMENT OF CONTAMINATION RISKS

A1.1 Flux Vector

The expression of flux vector is given in Equation (5),

$$\mathbf{K} = -D \operatorname{grad} c + \mathbf{v} \cdot c \qquad (5)/(A.1)$$

and in one-dimensional case, x-direction, the expression is

$$K_x = -D\frac{\partial c}{\partial x} + v_x \cdot c \qquad (A.2)$$

The numerical value of $\mathbf{K}$ gives the number of particles, which pass a supposed unit area, placed perpendicular of the direction of particle flow, per unit time.

A1.2 Unidirectional Air Flow

The expression of the concentration is given in Equation (2) for a continuous point source in a parallel flow with constant velocity, v_o, in x-direction. With the point source situated in the origin, the expression becomes:

$$c = \frac{q}{4\pi Dx} \cdot e^{-\frac{v_o(y^2 + z^2)}{4Dx}} \qquad (2)/(A.3)$$

The dispersion pattern in the x, y-plane ($z=0$) is schematically shown in Figure A1.

The expression for particle flux in the x-direction becomes:

$$K_x = \frac{q}{4\pi Dx} \cdot e^{-\frac{v_o(y^2 + z^2)}{4Dx}} \left(\frac{D}{x} - \frac{v_o(y^2 + z^2)}{4x^2} + v_o\right) \qquad (A.4.a)$$

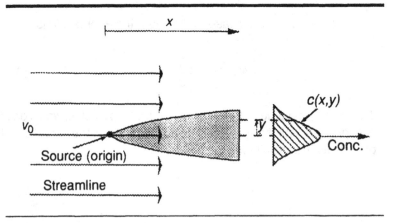

Figure A1. Schematically, dispersion pattern in the x,y-plane in a unidirectional air flow.

The expression for K_x in the streamline through the point source ($y=z=0$) is

$$K_x = \frac{q}{4\pi Dx} \left(\frac{D}{x} + v_0\right) \qquad \text{(A.4.b)}$$

With the numerical data given in Figure 1 (v_0=45 cm/s and D=2.4 cm²/s) it can easily be shown that at distances larger than 0.1 m K_x is proportional to air velocity, v_0, and concentration, c (error <1%).

The expressions in these cases become:

$$K_x = \frac{q \cdot v_0}{4\pi Dx} e^{-\frac{v_0 (y^2 + z^2)}{4Dx}} = v_0 \cdot c \qquad \text{(A.5.a)}$$

and

$$K_x = \frac{q \cdot v_0}{4\pi Dx} \qquad \text{(A.5.b)}$$

The particle flux in the direction perpendicular to the streamline that passes through the point source depends only on the diffusion. The expressions in y-direction and z-direction become:

$$K_y = -D\frac{\partial c}{\partial y} = \frac{yv_0}{2x} \cdot \frac{q}{4\pi Dx} e^{-\frac{v_0 (y^2 + z^2)}{4Dx}}$$

$$= \frac{y}{2x} v_0 \cdot c \qquad \text{(A.6.a)}$$

$$K_z = -D\frac{\partial c}{\partial z} = \frac{z}{2x} v_0 \cdot c \qquad \text{(A.6.b)}$$

and general in the y,z-plane

$$K_{p1} = \frac{(y^2 + z^2)^{1/2}}{2x} v_0 \cdot c \tag{A.7}$$

By using the data in Figure 1, it can be shown that in the critical region K_{p1} is up to 6% of K_x in Equation (A.5.a) and about 1% of K_x in Equation (A.5.b), except close to the streamline through the point source ($K_{p1} \approx 0$).

In a horizontal unidirectional air flow the settling velocities of bacteria carrying particles cannot always be neglected, and the particle flux due to gravitation will be proportional to the settling velocity and concentration ($v_s \cdot c$). According to Whyte [8] the average size of bacteria-carrying particles is 12 µm, which gives a settling velocity of 0.462 cm/s. This value is close to 1% of the air velocity in the unidirectional flow, i.e., the particle flux due to gravitation will be about one percent of K_x.

In conclusion, in a horizontal turbulent unidirectional air flow with particles of an average size of 12 µm, the particle flux caused by diffusion will approximately be in the same range as the particle flux due to gravitation, but much less than the the particle flux in the main flow direction.

A1.3 Vortex

The expression of the concentration in a vortex where the rotation takes place about the z-axis with the angular velocity ω and that source is located at the point (a, 0, 0) with the outward flow q is given with cylindrical coordinates in Equation (3).

$$c = \frac{q}{8\pi^{3/2}} \cdot \int_0^\infty e^{-(\rho^2 + a^2 - 2a\rho\cos(\varphi - \omega t) + z^2)\frac{1}{4Dt}} \frac{dt}{(Dt)^{3/2}} \tag{3)/(A.8}$$

With the point source in the center of the vortex (in origin) the expression in ρ, φ-plane (x,y-plane) becomes ($a=z=0$)

$$c = \frac{q}{8\pi^{3/2}} \cdot \int_0^\infty e^{-\frac{\rho^2}{4Dt}} \frac{dt}{(Dt)^{3/2}} = \frac{q}{4\pi D\rho} \qquad (A.9)$$

Along, for instance, the z-axis the concentration is

$$c = \frac{q}{4\pi Dz} \qquad (A.10)$$

The expression for particle flux in the ρ,φ-plane, when the air rotates with the angular velocity ω is

$$K\rho = -D\frac{\partial c}{\partial \rho} = \frac{q}{4\pi} \cdot \frac{1}{\rho^2} \qquad (A.11)$$

$$K\varphi = \omega\rho \cdot c = \frac{q}{4\pi} \cdot \frac{\omega}{D} \qquad (A.12)$$

The particle flux along the z-axis is:

$$K_z = -D\frac{\partial c}{\partial z} = \frac{q}{4\pi} \cdot \frac{1}{z^2} \qquad (A.13)$$

A certain preference should be given to the ρ, φ-plane (x,y-plane), as the contamination risk in this plane is higher than that of the direction along the z-axis.

Equations (A.11) and (A.12) show that $K\rho$ is greater than $K\varphi$ in a region close to the point source ($\rho < (D/\omega)^{1/2}$). Outside this region $K\rho$ is less than $K\varphi$, which has a constant value over the whole ρ,φ-plane (x,y-plane) and is proportional to the tangential velocity and the concentration.

Data from an experimental investigation of a macroscopic vortex (Ljungqvist [4]) show that when ρ is about 0.1 m $K\rho$ becomes equal to $K\varphi$, and at distances over 0.5 m from the point source, $K\rho$ is much less than $K\varphi$ (less than 5%).

In order to examine more closely the variation of the mean value with the distance from the central axis for an eccentrically situated point of emission, an averaging over the angle φ in Equation (3) is performed (integration and division with 2π) with consideration of the conditions in the ρ,φ-plane (x,y-plane) alone, i.e., $z = 0$. In an analogy with the approach in Equation (4) the expression of the concentration can be interpreted as a non-dimensional factor, which denotes F_1. Thus the expression of the mean value of the concentration becomes

$$c_m = \frac{q}{4\pi Da} \cdot F_1 \qquad\qquad (A.14)$$

where c_m = mean value of concentration over a closed streamline (circle)

         F_1 = non-dimensional factor, defined in Equation (A.15)

$$F_1 = \frac{2}{\pi(1 + \rho / a)} K(\alpha) \qquad\qquad (A.15)$$

where $K(\alpha)$ = the complete elliptical integral of the first kind.

$$\alpha = \arcsin \frac{2}{(1 + \rho / a)} \left(\frac{\rho}{a}\right)^{1/2}$$

Numerical values of F_1 as a function of ρ/a have been calculated and the results are shown in Figure A2.

Figure A2 shows with all desirable clarity that the mean value of
the concentration, over the entire region inside the point of emission,
is considerably higher than that of the outside. The above therefore
allows us to use the concept of contamination accumulation in the
context of vortices.

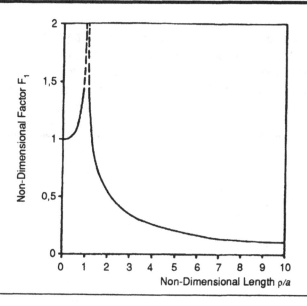

Figure A2. Mean value of concentration over a closed streamline
(circle) expressed in relative terms $0 \le \rho/a \le 10$.

A1.4 Completely Turbulent Mixing Air and Still Air

In a chamber with a height of H and an initial concentration c_o of
uniformly distributed monodisperse particles with a settling velocity
v_s, the expression for particle flux at a time t in the case with
completely mixing air is given in Equation (7):

$$K_s = v_s \cdot c_o \cdot e^{-\frac{v_s \cdot t}{H}} \qquad \text{(7)/(A.16)}$$

The expression for the case with still air is

$$K_s = v_s \cdot c_o \qquad \text{(A.17)}$$

The concentration in the case with completely turbulent mixing air (Equation (A.16)) reaches $1/e$ (ca. 37%) of the original concentration in the same time (H/v_s) that is required for the complete removal in the case with no motion of the air. Figure A3 shows schematically the particle flux for the two idealized situations:

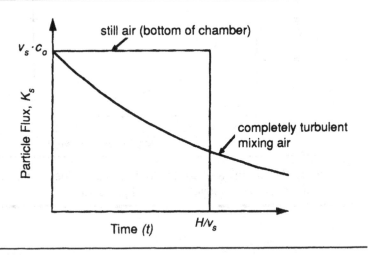

Figure A3. Particle flux for monodisperse particles for still air and completely turbulent mixing air.

The expression for the number of particles deposited is given in Equation (9) when the concentration of bacteria-carrying particles is uniformly distributed and constant during exposure time.

$$N_d = v_s \cdot c_b \cdot A_e \cdot t_e \qquad \text{(9)/(A.18)}$$

With the assumption, according to Whyte [8], that the average size of bacteria-carrying particles is 12 μm, the settling velocity will be 0.462 cm/s. The concentration in bacteria-carrying particles will yield the expression

$$c_b = \frac{N_d}{0.462 \cdot A_e \cdot t_e} \qquad (A.19)$$

With a settle plate of 90 mm and 4 hours exposure time the concentration in Equation (A.19) expressed in CFU/m^3 becomes

$$c_b = 2.36 \, N_d \approx 2.4 \, N_d \qquad (A.20)$$

In European Union draft Annex, *Manufacture of Sterile Medicinal Products* (1995), the following guidance values for microbiological monitoring of clean rooms in operation are given. The values for settle plates have been compared with the aid of Equation (A.20).

The calculated settle plate values obtained by Equation (A.20) are 20% higher than those of the air sample values. In consideration, as well the assumption in theory as the real case situations, the results should be considered within acceptable limits.

Table A.1. Guidance Values for Microbiological Monoriting of Clean Rooms

GRADE	Maximum number of viable organisms		
	Air sample CFU/m^3	Settle plate (90 mm) CFU/4 hours	Equation (A.20) CFU/m^3
A	<1	<1	—
B	10	5	12
C	100	50	120
D	200	100	240

Appendix 2

SIMULATION OF AIR FLOWS AND DISPERSION OF CONTAMINANTS THROUGH DOORWAYS IN A SUITE OF CLEAN ROOMS

The computer program used to simulate air flows and dispersion of contaminants through doorways in a suite of clean rooms, is called IDA, and it uses the Neutral Model Format (NMF). NMF models could be automatically translated into the format of a number of environments, such as TRNSYS, HVACSIM+, and SPARK. Based on NMF, independent libraries can be established. ASHRAE (American Society of Heating Refrigeration and Air Conditioning Engineers) has assumed the responsibility of approving any NMF changes. For a more thorough description, see Sahlin et al. [37, 38]. The simulation runs described in this appendix are partly reported by Isfält et al. [39], and they are made by Dr. Engelbrekt Isfält, Assoc. Professor at Building Services Engineering, KTH, Stockholm, Sweden.

The book edited by Whyte [36] shows an example of a typical suite of clean rooms configured to meet the requirements of producing a terminally sterilized injectable product. The computer drawing of this suite of clean rooms is shown in Figure A4 and arrows between the rooms indicate air flows through pressure stabilizer dampers, air leakages through closed doors, or air flows through door openings.

In the simulation runs, the following basic data are given for the different types of rooms.

Clean Filling Room (called fill room)

Air changes	20/h
Pressure difference	+45 Pa
Temperature	+20°C

Preparation Areas

Equipment and component preparation area (called prep room) and solution preparation area (called prep prod).

Air changes	20/h
Pressure difference	+30 Pa
Temperature	+20°C

Clean Changing Area (called locker room)

No mechanical ventilation.
Flow through pressure stabilization dampers.

Pressure difference	about +15 Pa
Temperature	+20°C

Entry Airlocks

Component entry airlock (called lock 1) and material entry airlock (called lock 2)

No mechanical ventilation.
Flow through pressure stabilization dampers.

Pressure differences   about + 15 Pa

Temperature   +20°C

Figure A5 shows obtained pressure differences for every room when doors are closed, supply/exhaust air in air changes per hour (box labels), and leakages between zones. It can be seen that rooms with mechanical ventilation get expected pressure differences and rooms with only overflow through dampers receive higher or lower values than expected.

Figures A6–A10 show the pressure differences in a suite of rooms with one or several doors opened.

Figures A11–A22 show the air flows through doorways and contamination levels in rooms when temperature differences occur between rooms. In the simulation runs, the contamination source has been situated either in the locker room or prep room and doors with width 1 m and height 2 m have been open one at a time, fully opened (Figures A11–A18), and approximately half opened (Figures A19–A22).

To take into account all different parameters needed to theoretically describe the situations discussed here, close to three hundred equations are required. The results from the simulation runs have not yet been completely verified with measurements in a real case situation with a suite of clean rooms. Due to this, a number of constants in the simulation model have been chosen from literature instead of determined from experimental data. In spite of this, certain trends can be discerned.

From Figures A11, A13, and A16 for a fully opened door (area 1 × 2 m²) and in Figures A19 and A21 for a half-opened door, it can be estimated that the air flows reach the same ranges that are given by using Figure 38. When the contamination source is situated in a room with lower pressure difference than that of ambient/adjacent rooms (doors closed), there will be—when the door is opened—no contamination dispersion at temperature differences of zero, but a significant change will occur at small temperature differences (2–4°C).

One important as well as a practical conclusion should be drawn: When temperature differences occur between clean rooms with air changes and pressure differences according to present regulatory demands, there will always be a risk of contamination ingress when a door is opened.

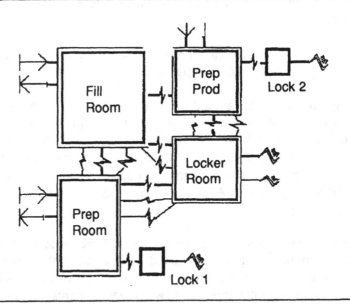

Figure A4. Computer drawing of a typical suite of clean rooms for terminally sterilized injectables.

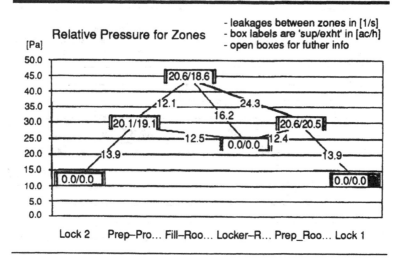

Figure A5. Pressure differences, supply/exhaust air in air changes per hour, and leakages between zones in a suite of clean rooms with closed doors.

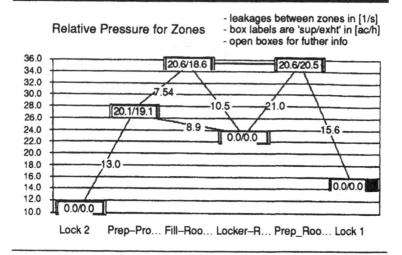

Figure A6. Pressure differences in a suite of clean rooms when doors between prep room and fill room is open.

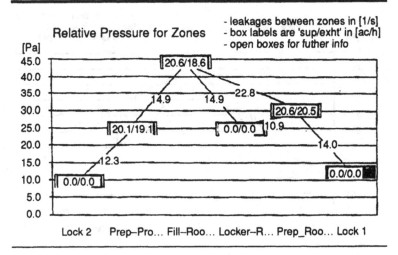

Figure A7. Pressure differences in a suite of clean rooms, when door between prep prod and locker room is open.

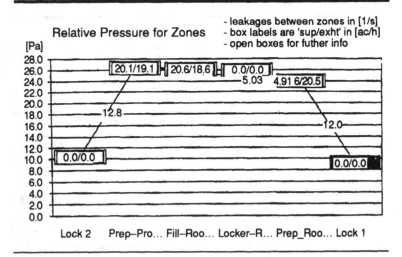

Figure A8. Pressure differences in a suite of clean rooms, when door between locker room and fill room is open.

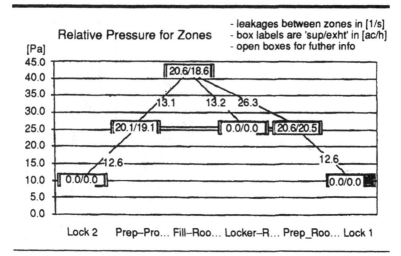

Figure A9. Pressure differences in a suite of clean rooms, when doors between prep prod and locker room, and between prep room and locker room are open.

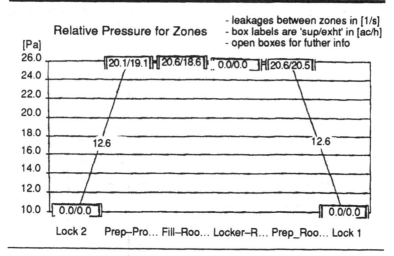

Figure A10. Pressure differences in a suite of clean rooms, when doors between prep prod and fill room, and between prep room and fill room are open.

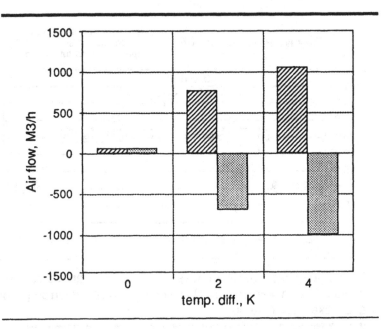

Figure A11. Air flow through doorway when door between locker room and prep room is fully opened (area 1×2 m²). Increasing temperature in prep room.

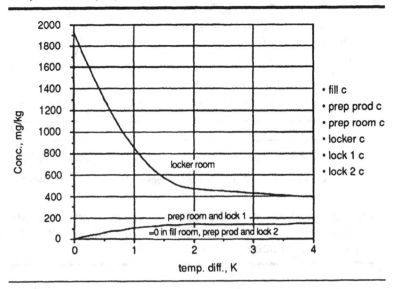

Figure A12. Contamination levels in a suite of clean rooms when door between locker room and prep room is open according to Figure A11. Contamination source in locker room.

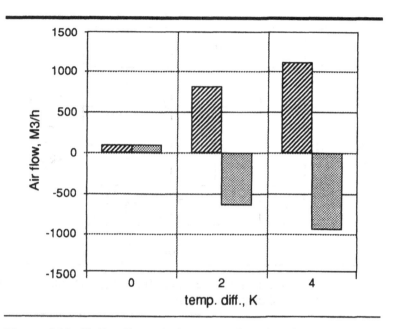

Figure A13. Air flow through doorway, when door between locker room and fill room is fully opened (area 1 × 2 m²). Increasing temperature in fill room.

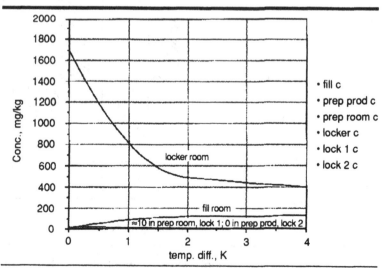

Figure A14. Contamination levels in a suite of clean rooms, when door between locker room and fill room is open according to Figure A13. Contamination source in locker room.

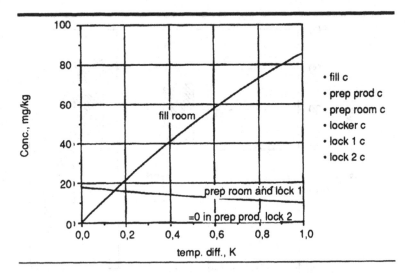

Figure A15. Contamination levels with enlarged scales (using the same situation shown in Figure A14).

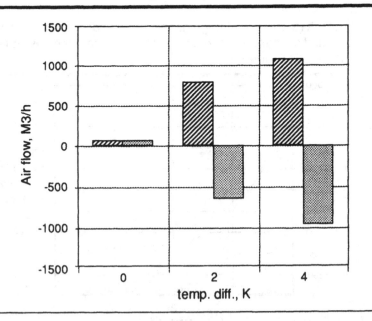

Figure A16. Air flow through doorway, when door between prep room and fill room is fully opened (area 1 × 2 m²). Increasing temperature in fill room.

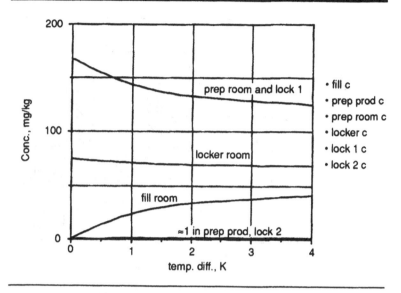

Figure A17. Contamination levels in a suite of clean rooms when door between prep room and fill room is open according to Figure A16. Contamination source in prep room.

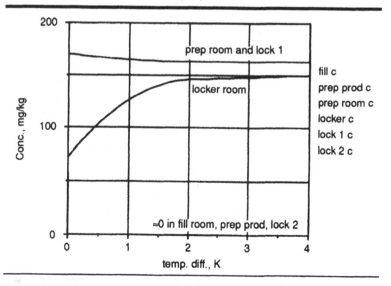

Figure A18. Contamination levels in a suite of clean rooms, when door between locker room and prep room is open according to Figure A11. Contamination source in prep room.

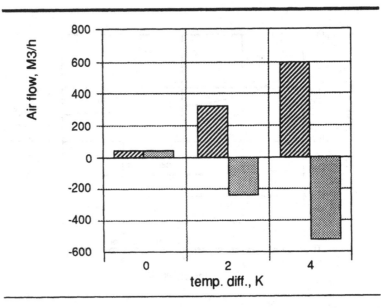

Figure A19. Air flow through doorway, when door between locker room and prep room is half-opened. Increasing temperature in prep room.

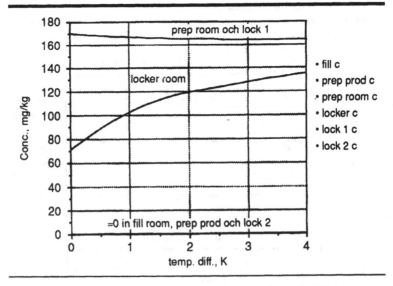

Figure A20. Contamination levels in a suite of clean rooms, when door between locker room and prep room is open according to Figure A19. Contamination source in prep room.

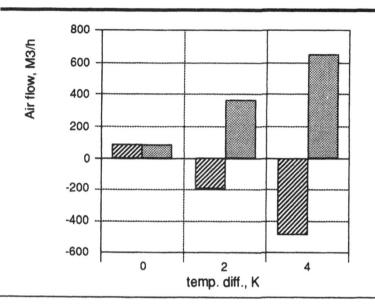

Figure A21. Air flow through doorway, when door between prep room, and fill room is half-opened. Increasing temperature in prep room.

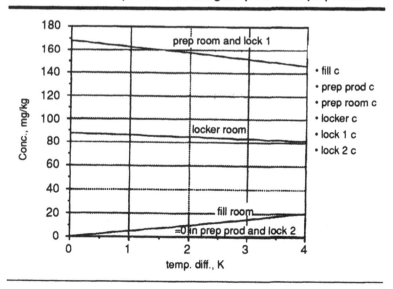

Figure A22. Contamination levels in a suite of clean rooms, when door between prep room and fill room is open according to Figure A21. Contamination source in prep room.

Appendix 3

REGULATORY REQUIREMENTS

A3.1 U.S. CGMP Regulations (21 CFR 211)

Excerpts from Part 211, "Current good manufacturing practice for finished pharmaceuticals" (4-1-96 Edition).

Subpart C Buildings and Facilities

§ 211.42   Design and construction features.

(a) Any building or buildings used in the manufacture, processing, packing or holding of drug product shall be of suitable size, construction and location to facilitate cleaning, maintenance, and proper operations.

(b) Any such building shall have adequate space for the orderly placement of equipment and materials to prevent mixups between different components, drug product containers, closures, labeling, in-process materials, or drug products, and to prevent contamination. The flow of components, drug product containers, closures, labeling, in-process materials, and drug products through the building or buildings shall be designed to prevent contamination.

(c) Operations shall be performed within specifically defined areas of adequate size. There shall be separate or defined areas or such other control systems as are necessary for the firm's operations to prevent contamination or mixup during the course of the following procedures:

(1)   Receipt, identification, storage, and withholding from use of components, drug product containers, closures, and labeling, pending the appropriate sampling, testing, or examination by the quality control unit before release for manufacturing or packaging;

(2)   Holding rejected components, drug product containers, closures, and labeling before disposition;

(3)   Storage of released components, drug product containers, closures, and labeling;

(4)   Storage of in-process materials;

(5)   Manufacturing and processing operations;

(6)   Packaging and labeling operations;

(7)   Quarantine storage before release of drug products;

(8)   Storage of drug products after release;

(9)   Control and laboratory operations;

(10) Aseptic processing, which includes as appropriate:

(i)   Floors, walls, and ceilings of smooth, hard surfaces that are easily cleanable;

(ii)   Temperature and humidity controls;

(iii) An air supply filtered through high-efficiency particulate air filters under positive pressure, regardless of whether flow is laminar or nonlaminar;

(iv) A system for monitoring environmental conditions;

(v) A system for cleaning and disinfecting the room and equipment to produce aseptic conditions;

(vi) A system for maintaining any equipment used to control the aseptic conditions.

(d) Operations relating to the manufacture, processing, and packing of penicillin shall be performed in facilities separate from those used for other drug products for human use.

[43 FR 45077, Sept. 29, 1978, as amended at 60 FR 4091, Jan. 20, 1995.]

§ 211.44   Lighting.

Adequate lighting shall be provided in all areas.

§ 211.46   Ventilation, air filtration, air heating and cooling.

(a) Adequate ventilation shall be provided.

(b) Equipment for adequate control of air pressure, micro-organisms, dust humidity, and temperature shall be provided when appropriate for the manufacture, processing, packing, or holding of a drug product.

(c) Air filtration systems, including prefilters and particulate matter air filters, shall be used when appropriate on air supplies to production areas. If air is recirculated to production areas, measures shall be taken to control recirculation of dust from production. In areas where air contamination occurs during production, there shall be adequate exhaust systems or other systems adequate to control contaminants.

(d) Air-handling systems for the manufacture, processing, and packing of penicillin shall be completely separate from those for other drug products for human use.

§ 211.48   Plumbing.

(a) Potable water shall be supplied under continuous positive pressure in a plumbing system free of defects that could contribute contamination to any drug product. Potable water shall meet the standards prescribed in the Environmental Protection Agency's Primary Drinking Water Regulations set forth in 40 CFR Part 141. Water not meeting such standards shall not be permitted in the potable water system.

(b)   Drains shall be of adequate size and, where connected directly to a sewer, shall be provided with an air break or other mechanical device to prevent back-siphonage.

[43 FR 45077, Sept. 29, 1978, as amended at 48 FR 11426, Mar. 18, 1983.]

§ 211.50   Sewage and refuse.

Sewage, trash, and other refuse in and from the building and immediate premises shall be disposed of in a safe and sanitary manner.

§ 211.52   Washing and toilet facilities.

Adequate washing facilities shall be provided, including hot and cold water, soap or detergent, air driers or single-service towels, and clean toilet facilities easily accessible to working areas.

§ 211.56   Sanitation.

(a) Any building used in the manufacture, processing, packing, or holding of a drug product shall be maintained in a clean and sanitary condition. Any such building shall be free of infestation by rodents, birds, insects, or other vermin (other than laboratory animals). Trash and organic waste matter shall be held and disposed of in a timely and sanitary manner.

(b) There shall be written procedures assigning responsibility for sanitation and describing in sufficient detail the cleaning schedules, methods, equipment, and materials to be used in cleaning the buildings and facilities; such written procedures shall be followed.

(c) There shall be written procedures for use of suitable rodenticides, insecticides, fungicides, fumigating agents, and cleaning and sanitizing agents. Such written procedures shall be designed to prevent the contamination of equipment, components, drug product containers, closures, packaging, labeling materials, or drug products and shall be followed. Rodenticides, insecticides, and fungicides shall not be used unless registered and used in accordance with the Federal Insecticide, Fungicide, and Rodenticide Act (7 U.S.C.135).

(d) Sanitation procedures shall apply to work performed by contractors or temporary employees as well as work performed by full-time employees during the ordinary course of operations.

§ 211.58   Maintenance.

Any building used in the manufacture, processing, packing, or holding of a drug product shall be maintained in a good state of repair.

A3.2   Excerpts from the FDA document, "Guideline on Sterile Drug Products Produced by Aseptic Processing" (June 1987)

III. Building and facilities

Requirements

Section 211.42 (design and construction features) requires, in part, that there be separate or defined areas of operation to prevent contamination, and that for aseptic processing there be, as appropriate, an air supply filtered through high efficiency particulate air (HEPA) filters under positive pressure, and systems for monitoring the environment and maintaining equipment used to control aseptic conditions.

Section 211.46 (ventilation, air filtration, air heating and cooling) requires, in part, that equipment for adequate control over air pressure, microorganisms, dust, humidity, and temperature be provided where appropriate and that air filtration systems, including prefilters and particulate matter air filters, be used when appropriate on air supplies to production areas.

Guidance

In aseptic processing there are various areas of operation which require separation and control, with each area needing different degrees of air quality depending on the nature of the operation. Two exposure areas are of particular importance to drug product quality—critical areas and controlled areas.

CRITICAL AREAS

A critical area is one in which the sterilized dosage form, containers, and closures are exposed to the environment. Activities that are conducted in this area include manipulations of these sterilized materials/product prior to and during filling/closing operations. These operations are conducted in what is typically called the "aseptic core" or "aseptic processing" area.

This area is critical because the product is not processed further in its immediate container and is vulnerable to contamination. Therefore, in order to maintain the quality and, specifically, the sterility of the product, the environment in the immediate proximity of the actual operations should be of the highest quality.

One aspect of environmental quality is the particulate content of the air. Particulates are significant because they may enter a product and contaminate it physically or, by acting as a vehicle for microorganisms, biologically. It is therefore important to minimize the particle content of the air and to effectively remove those particles which are present. Air in the immediate proximity of exposed sterilized containers/closures and filling/closing operations is of acceptable particulate quality when it has a per-cubic-foot particle count of no more than 100 in a size range of 0.5 micron an larger (Class 100) when measured not more than one foot away from the work site, and upstream of the air flow, during filling/closing operations. The agency recognizes that some powder filling operations may generate high levels of powder particulates which, by their nature, do not pose a risk of product contamination. It may not, in these cases, be feasible to measure air quality within the one foot distance and still differentiate "background noise" levels of powder particles from air contaminants which can impeach product quality. In these instances, it is nonetheless important to sample the air in a manner which, to the extent possible, characterizes the true level of extrinsic particulate contamination to which the product is exposed.

Air in critical areas should be supplied at the point of use as HEPA filtered laminar flow air, having a velocity sufficient to sweep particulate matter away from the filling/closure area. Normally, a velocity of 90 feet per minute, plus or minus 20%, is adequate, (Refs. 1 and 2) although higher velocities may be needed where the operations generate high levels of particulates or where equipment configuration disrupts laminar flow.

Air should also be of high microbial quality. An incidence of no more than one colony forming unit per 10 cubic feet is considered as attainable and desirable (Ref. 3).

Air is not the only gas in the proximity of filling/closing operations which should be of high particulate and microbial quality. Other gases, such as nitrogen or carbon dioxide, which contact the product, container/closure, or product contact surfaces, e.g., purging or overlaying, should be sterile filtered. In addition, compressed air should be free from demonstrable oil vapors.

Critical areas should have a positive pressure differential relative to adjacent less clean areas, a pressure differential of 0.05 inch of water is acceptable.

CONTROLLED AREAS

The controlled area, the second type of area in which it is important to control the environment, is the area where unsterilized product, in-process materials, and container/closures are prepared. This includes areas where components are compounded, and where components, in-process materials, drug products and drug product contact surfaces of equipment, containers, and closures, after final rinse of such surfaces, are exposed to the plant environment. This environment should be of a high microbial and particulate quality in order to minimize the level of particulate contaminants in the final product and to control the microbiological content (bioburden) of articles and components which are subsequently sterilized.

Air in the controlled areas is generally of acceptable particulate quality if it has a per-cubic-foot particle count of not more than 100,000 in a size range of 0.5 micron and larger (Class 100,000) when measured in the vicinity of the exposed articles during periods of activity. With regard to microbial quality, an incidence of no more than 25 colony forming units per 10 cubic feet is acceptable (Ref. 3)

In order to maintain air quality in controlled areas, it is important to achieve a sufficient air flow and a positive pressure differential relative to adjacent uncontrolled areas. In this regard, an air flow sufficient to achieve at least 20 air changes per hour and, in general, a pressure differential of at least 0.05 inch of water (with all doors closed), are acceptable. When doors are open, outward airflow should be sufficient to minimize ingress of contamination.

Gases other than ambient air may also be used in controlled areas. Such gases should, if vented to the area, be of the same quality as ambient air. Compressed air should be free from demonstrable oil vapor.

In addition to these production areas, there may be certain pieces of equipment which should be supplied with high quality filtered air. This is especially important where the air in the equipment will contact sterilized material or material which should have a low microbial or particulate content. For example, bacterial retentive filters should be used for lyophilizer vacuum breaks and hot air sterilizer vents to ensure that air coming in contact with a sterilized product is sterile. Likewise, air admitted to unpressurized vessels containing sterilized liquid should also be filtered. Air in tanks used to hold material which must be of a high microbial quality should be filtered too, and the filters should be dry to prevent wetting by condensation with subsequent blockage or microbial grow-through (two ways of achieving this are providing heat to the filter and use of hydrophobic filters). It is important that these equipment air filters be periodically integrity tested.

An acceptable system for maintaining air quality includes testing HEPA filters for integrity. Integrity testing should be performed initially when the units are first installed in order to detect leaks around the sealing gaskets, through the frames or through the filter media. Thereafter, integrity tests should be performed at suitable intervals. Usually it is sufficient to perform such testing at least twice a year for critical areas; however, more frequent testing may be needed when air quality is found to be unacceptable low or as part of an investigation into a finding of non-sterility in a drug product.

One acceptable method of testing the integrity of HEPA filters is use of a dioctylphtalate (DOP) aerosol challenge. Inasmuch as a HEPA filter is one capable of retaining 99.97 percent of particulates greater than 0.3 micron in diameter, it is important to assure that whatever substance is used as a challenge will have a sufficient number of particles of this size range. An acceptable DOP challenge involves introducing a DOP aerosol upstream of the filter in a concentration of 80 to 100 micrograms/liter of air at the filter's designed airflow rating and then scanning the downstream side of the filter with an appropriate photometer probe at a sampling rate of at least one cubic foot per minute. The probe should scan the entire filter face and frame at a position about one to two inches from the face of the filter (Ref. 1). A single probe reading equivalent to 0.01 percent of the upstream challenge is considered as indicative of a significant leak which should be repaired.

Use of particle counters without introducing particles of known size upstream of the filter is ineffective for detecting leaks.

The reader should note that there is a difference between filter integrity testing and efficiency testing. Integrity testing is performed to detect leaks from the filter media, filter frame and seal. The challenge is a polydispersed aerosol usually composed of particles ranging in size from one to three microns. The test is done in place and the filter face is scanned with a probe; the measured downstream leakage is taken as a percent of the upstream challenge. The efficiency test, on the other hand, is used to determine the filter's rating. The test uses a monodispersed aerosol of 0.3 micron size particles, relates to filter media, and usually requires special equipment. Downstream readings represent an average over the entire filter surface. Therefore, leaks in a filter may not be detected by an efficiency test.

It is also important to monitor air flow velocities for each HEPA filter according to a program of established intervals because significant reductions in velocity can increase the possibility of contamination and changes in velocity can affect the laminarity of the airflow. Airflow patterns should be tested for turbulence that would interfere with the sweeping action of the air.

REFERENCES

1. Federal Standard 209 B, Clean Room and Work Station Requirements, Controlled Environment, April 24, 1973

2. Technical Order 00-25-203, Contamination Control of Aerospace Facilities, U.S. Air Force, December 1, 1972.

3. NASA Standard for Clean Room and Work Stations for Microbially Controlled Environment, Publication NHB 5340.2 (August 1967)

A3.3 Excerpts from the EC GMP Draft Annex, "Manufacture of Sterile Medicinal Products"

Examples of operations to be carried out in the various grades are given in the table below

Grade	Examples of operations
A	Aseptic preparation and filling. Filling of products to be terminally sterilised when products are unusually at risk.
B	Transfer and storage of containers of freeze-dried products and components for aseptic filling.
C	Preparation of solutions and components for subsequent sterile filtration and aseptic filling. Preparation of solutions and components for subsequent filling and terminal sterilisation when products or components are considerably exposed or unusually at risk.
	Filling of products to be terminally sterilized.
D	Preparation of solutions and components for subsequent filling and terminal sterilisation.

Clean areas for production of sterile products are classified according to the required characteristics of the environment. Each manufacturing operation requires an appropriate environmental cleanliness level in the operational state in order to minimise the risks of particulate or microbial contamination of the product or materials being handled.

In order to meet "in operation" conditions these areas should be designed to reach certain specified air-cleanliness levels in the "at rest" occupancy state. During the "at rest" state an area is fully equipped with the ventilation systems working but unmanned. In the state "in operation" state the clean room is manned with the normal number of personnel present.

For the manufacture of sterile medicinal products normally 4 grades can be distinguished.

Grade A: The local zone for high risk operations, e.g. filling zone, stopper bowls, open ampoules and vials, aseptic connections. Normally such conditions are provided by a laminar air flow work station. Laminar air flow systems should provide an homogeneous air speed of 0.30 m/s for vertical flow and 0.45 m/s for horizontal flow.

Grade B: In case of aseptic preparation and filling the background environment for grade A zone. Grade C and D: Clean areas for carrying out less critical stages in the manufacture of sterile products.

The particulate conditions given in the table for the "at rest" state should be achieved after throughout the environment where unmanned, and recovered after a short "clean up" period, usually between 15–20 minutes. The particulate condition for grade A in operation given in the table should be maintained in the zone immediately surrounding the product whenever the product or open container is exposed to the environment. It is accepted that it may not always be possible to demonstrate conformity with particulate standards at the point of fill when filling is in progress, due to the generation of particles or droplets from the product itself.

In order to control the microbiological and particulate cleanliness of the various grades in operation, the areas should be monitored. A variety of methods should be used for example volumetric air sampling, settle plates, surface sampling (swabs, contact plates).

Where aseptic operations are performed monitoring should be frequent and methods should include settle plates, volumetric air sampling and surface sampling. Monitoring results should be considered when reviewing batch documentation for finished product release. Air should be monitored in the "in operation" state. Critical surfaces and personnel should be monitored immediate after operation.

Additional microbiological monitoring is also required outside production operations, e.g. after validation of systems, cleaning and fumigation.

	at rest		in operation (c)	
GRADE	maximum permitted number of particles/m^3 equal to or above			
	0,5 μm	5 μm	0,5 μm	5 μm
A	3 500	-	3,500	-
B(a)	3 500	-	350 000	2 000
C(a)	350 000	2 000	3 500 000	20 000
D(a)	3 500 000	20 000	-	-

Notes

(a) In order to reach the B, C and D air grades, the number of air changes should generally be higher than 20 per hour in a room with a good air flow pattern and appropriate filters; HEPA for grades A, B and C.

(b) Appropriate alert and action limits should be set for the particular operation.

The guidance given for the maximum permitted number of particles in the "at rest" condition corresponds approximately to the US Federal Standard 209 E as follows: Class 100 (grades A and B). Class 10000 (grade C) and Class 100000 (grade D).

(c) Recommended limits for contamination may be exceeded on isolated occasions and require only an examination of the production conditions and the control system. If the frequency is high or shows an upward trend then action should be taken.

Guidance values for microbiological monitoring of clean rooms in operation.

GRADE	Maximum number of viable organisms (a)			
	air sample cfu/m^3	settle plate (90mm) cfu/4 hour	contact plate (55 mm) cfu	glove print 5 fingers cfu
A	<1 (b)	<1 (b)	<1 (b)	<1 (b)
B	10	5	5	5
C	100	50 (c)	25	-
D	200	100 (c)	50	-

Notes

(a) Recommended limits for contamination may be exceeded on isolated occasions and require only an examination of the production conditions and the control system. If the frequency is high or shows an upward trend then action should be taken.

(b) Low values involved here are only reliable when a larger number of samples is taken.

(c) For Grades C and D settle plates may be exposed for less than 4 hours.

A3.4 <1116> Microbiological Evaluation of Clean Rooms and Other Controlled Environments

United States Pharmacopeial Convention, Inc.

Table 3. Air Cleanliness Levels in Colony-Forming Units (cfu) in Controlled Environments (Using a Slit-to-Agar Sampler or Equivalent)

	Class*	cfu per cubic meter of air**	cfu per cubic feet of air
S1	U.S. Customary		
M3.5	100	Less than 3	Less than 0.1
M5.5	10,000	Less than 20	Less than 0.5
M6.5	100,000	Less than 100	Less than 2.5

* As defined in Federal Standard 209E, September 1992.

** A sufficient volume of air should be sampled to yield finite results.

INDEX

Printed in the United States
by Baker & Taylor Publisher Services